Lecture Notes in Mathematics

Edited by A. Dold and B. Eckmann

Subseries: Scuola Normale Superiore, Pisa
Adviser: E. Vesentini

1164

Mauro Meschiari
John H. Rawnsley
Simon Salamon

Geometry Seminar "Luigi Bianchi" II – 1984

Lectures given at the Scuola Normale Superiore

Edited by E. Vesentini

Springer-Verlag
Berlin Heidelberg New York Tokyo

Authors

Mauro Meschiari
Dipartimento di Matematica Pura ed Applicata, Università di Modena
Via G. Campi 213/B, 41100 Modena, Italy

John H. Rawnsley
Mathematics Institute, University of Warwick
Coventry CV4 7AL, England

Simon Salamon
Mathematical Institute, University of Oxford
24–29 St. Giles', Oxford OX1 3LB, England

Editor

Edoardo Vesentini
Scuola Normale Superiore
Piazza dei Cavalieri 7, 56100 Pisa, Italy

Mathematics Subject Classification (1980): 15A21, 15A57, 32A07, 41D25; 53C15
53C21, 58G30; 53B, 53C

ISBN 3-540-16048-5 Springer-Verlag Berlin Heidelberg New York Tokyo
ISBN 0-387-16048-5 Springer-Verlag New York Heidelberg Berlin Tokyo

Printing and binding: Beltz Offsetdruck, Hemsbach/Bergstr.
2146/3140-543210

MAURO MESCHIARI

A CLASSIFICATION OF REAL AND COMPLEX FINITE DIMENSIONAL J*-ALGEBRAS

JOHN H. RAWNSLEY

f-STRUCTURES, f-TWISTOR SPACES AND HARMONIC MAPS

SIMON SALAMON

HARMONIC AND HOLOMORPHIC MAPS

CONTENTS

MAURO MESCHIARI: A CLASSIFICATION OF REAL AND COMPLEX FINITE DIMENSIONAL J*-ALGEBRAS

JOHN H. RAWNSLEY: f-STRUCTURES, f-TWISTOR SPACES AND HARMONIC MAPS

SIMON SALAMON: HARMONIC AND HOLOMORPHIC MAPS

MAURO MESCHIARI

A CLASSIFICATION OF REAL AND COMPLEX FINITE DIMENSIONAL J*-ALGEBRAS

PREFACE

The purpose of this paper is to give a complete classification of real and complex finite dimensional J*-algebras

First of all we consider real finite dimensional J*-algebras. We characterize twelve infinite families of real matrix J*-algebras and prove that any finite dimensional real J*-algebra is J*-isomorphic to a finite direct sum whose summands are chosen in the above families.

Next, using the above classification devised for real J*-algebras, we obtain a complete classification for finite dimensional complex J*-algebras. To this end we characterize four families of complex matrix J*-algebras and prove that any finite dimensional complex J*-algebra is J*-isomorphic (as a complex J*-algebra) to a finite direct sum whose summands belong to any of the above families.

The classification of finite dimensional complex J*-algebras obtained in this way happens to coincide with the classification given by S.A. Harris [2], in a completely independent way. At the same time we solve, negatively, the problem of the existence of a complex J*-algebra whose associated unitary disc is one of the two exceptional domains in the Cartan classification This same problem was solved independentely by O. Loss and K. McCrimmon [3].

For convenience we divide this paper in four parts:

Part one. Definitions and elementary properties are exposed. We give some relevant properties of some J*-algebras and prove that any finite dimensional J*-algebra is the direct sum of finite dimensional irreducible ones.

Part two. We give some example of matrix J*-algebras and determine which ones are J*-isomorphic.

Part three. First we classify some particular J*-algebras (J*-algebras of height 2), then we extend this classification to all finite dimensional real J*-algebras.

Part four. We study some properties characterizing complex finite dimensional indecomposable J*-algebras (considered as real J*-algebras)

Then, using the above properties, we sort out four finite families of real matrix J*-algebras which provide all the models for complex finite dimensional indecomposable J*-algebras. Finally we prove that the existence of a real J*-isomorphism between two complex J*-algebras implies the existence of a complex J*-isomorphism.

I thank warmly prof. Edoardo Vesentini whose patience in discussing this paper with me helped me greatly in attaining the solution.

PART 1.

In what follows $\mathbb{K}$ will be either the real number field $\mathbb{R}$ or the complex number field $\mathbb{C}$. Moreover, whenever V is an Hilbert vector space on the field $\mathbb{K}$, let $<,>$ $(<,>_{\mathbb{K}})$ denote the inner product in V and $|\ |$ the relative norm.

Let V and W be two Hilbert vector spaces on $\mathbb{K}$ and $f : V \to W$ a bounded $\mathbb{K}$-linear operator of V into W. $f^* : W \to V$ will be the $\mathbb{K}$-linear bounded operator defined by

$$<f(x),y>_{\mathbb{K}} = <x,f^*(y)>_{\mathbb{K}}, \text{ whenever } x \in V \text{ and } y \in W.$$

A J*-algebra over $\mathbb{K}$, or a $\mathbb{K}$-J*-algebra, is a closed $\mathbb{K}$-linear subspace j of the space $L(V,W)$, of all bounded $\mathbb{K}$-linear operators of the $\mathbb{K}$-Hilbert space V into the $\mathbb{K}$-Hilbert space W (endowed with the operator norm), such that

$$aa^*a \in j, \text{ whenever } a \in j.$$

From now on, J*-algebra will always mean finite dimensional J*-algebra, i.e. a J*-algebra of finite dimension as a vector space.

Let V be a complex Hilbert space. V is naturally a real Hilbert space too and the two inner products on V are bounded by the identities:

$$<x,y>_{\mathbb{R}} = \frac{1}{2}(<x,y>_{\mathbb{C}} + <y,x>_{\mathbb{C}}), \quad \text{for all } x,y \in V;$$

$$<x,y>_{\mathbb{C}} = <x,y>_{\mathbb{R}} + i<x,iy>_{\mathbb{R}}, \quad \text{for all } x \in V.$$

Hence, whenever $f : V \to W$ is a bounded linear operator, we have

$$<f(x),y>_{\mathbb{R}} = <x,f^*(y)>_{\mathbb{R}}, \text{ for all } x \in V \text{ and } y \in W,$$

and any $\mathbb{C}$-J*-algebra is naturally a $\mathbb{R}$-J*-algebra too.

Let $\mathcal{J}_1$ and $\mathcal{J}_2$ be two $\mathbb{K}$-J*-algebras; a bounded $\mathbb{K}$-isomorphism $L : \mathcal{J}_1 \to \mathcal{J}_2$ is a $\mathbb{K}$-J*-isomorphism if

$$L(aa^*a) = L(a)\, L(a)^*\, L(a), \quad \text{for all } a \in \mathcal{J}_1.$$

It is trivial to prove that any $\mathbb{C}$-J*-isomorphism of the $\mathbb{C}$-J*-algebra $\mathcal{J}_1$ onto the $\mathbb{C}$-J*-algebra $\mathcal{J}_2$ is a $\mathbb{R}$-J*-isomorphism of $\mathcal{J}_1$ onto $\mathcal{J}_2$ (considered as $\mathbb{R}$-J*-algebras) too.

In the following we shall omit the prefixes $\mathbb{K}$-, $\mathbb{R}$- and $\mathbb{C}$- whenever possible.

Given a,b, and c in any J*-algebra $\mathcal{J}$, then

$$\{abc\} = \frac{1}{6}(ab^*c + cb^*a + ac^*b + bc^*a + ba^*c + ca^*b)$$

is in $\mathcal{J}$.

<u>LEMMA 1</u>. Let $\mathcal{J}$ be a finite dimensional J*-algebra of bounded linear operators of the Hilbert space V into the Hilbert space W. For any $a \in \mathcal{J}$ there exist an orthonormal basis $\mathcal{B}$ of V and an orthonormal basis $\mathcal{R}$ of W such that for any $u \in \mathcal{B}$, $a(u)$ is proportional to an element of $\mathcal{B}$.

<u>PROOF</u>. Let $I_p(a^*a)$ denote the point spectrum of the bounded endomorphism $a^*a : V \to V$, i.e.

$$I_p(a^*a) = \{\lambda : \lambda \in \mathbb{K} \text{ such that } a^*a - \lambda 1_V \text{ is not injective}\}.$$

We shall prove that:

i) $I_p(a^*a) \subset \mathbb{R}$;

ii) $V_\lambda = \mathrm{Ker}(a^*a - \lambda 1_V) = \mathrm{Im}(a^*a - \lambda 1_V)^\perp \neq \{0\}$, whenever $\lambda \in I_p(a^*a)$; (if U is a subspace of V, $U^\perp$ is the orthogonal complement of U in V)

iii) V_λ and V_μ are orthogonal ($V_\lambda \perp V_\mu$) whenever $\lambda, \mu \in I_p(a^*a)$ with $\lambda \neq \mu$;

iv) $V = \Sigma_\lambda V_\lambda$ $(\lambda \in I_p(a^*a))$.

Let $\lambda \in I_p(a^*a)$ and $x \in \mathrm{Ker}(a^*a - \lambda 1_V) - \{0\}$; we have

$$\lambda |x|^2 = <\lambda x, x> = <a^*a(x), x> = <x, a^*a(x)> = <x, \lambda x> = \bar{\lambda}|x|^2$$

and hence i).

Let $\lambda \in I_p(a^*a)$, $x \in \mathrm{Ker}(a^*a - \lambda 1_V)$, $y \in \mathrm{Im}(a^*a - \lambda 1_V)^\perp$ and $z \in V$. We have

$$0 = <(a^*a - \lambda 1_V)(x), z> = <x, (a^*a - \lambda 1_V)(z)>$$

and

$$0 = <y, (a^*a - \lambda 1_V)(z)> = <(a^*a - \lambda 1_V)(y), z>.$$

Then ii) follows trivially.

Let $\lambda, \mu \in I_p(a^*a)$, with $\lambda \neq \mu$, $x \in \mathrm{Ker}(a^*a - \lambda 1_V)$ and $y \in \mathrm{Ker}(a^*a - \mu 1_V)$. We have

$$\lambda <x,y> = <\lambda x, y> = <a^*a(x), y> = <x, a^*a(y)> = <x, \mu y> = \mu <x,y>$$

and hence

$$(\lambda - \mu) <x,y> = 0.$$

which proves iii).

Assume now that $M = (\Sigma_\lambda V_\lambda)^\perp \neq \{0\}$ $(\lambda \in I_p(a^*a))$ and let $x \in M - \{0\}$, $p = a^*a - 2|a^*a|1_V$. Clearly $p : V \to V$ has a bounded inverse p^{-1} since the spectral radius of a^*a, $\rho(a^*a)$, is $|a^*a|$. Now let us consider the set $I = \{p^m(x) : m \in \mathbb{Z}\}$. We have

$$<p^m(x), y> = <x, p^m(y)> = (\lambda - 2|a^*a|)^m <x,y> = 0,$$

whenever $\lambda \in I_p(a^*a)$, $y \in V_\lambda$ and $m \in \mathbb{Z}$, and hence $I \subset M$.

Since j is finite dimensional, there exist a natural number r and n scalar $\lambda_0, \ldots, \lambda_{n-1}$ such that

$$a(a^*a)^n = \lambda_0 a + \lambda_1 a(a^*a) + \ldots + \lambda_{n-1} a(a^*a)^{n-1}$$

and hence

$$(a^*a)^{n+1} = \lambda_0 (a^*a) + \lambda_1 (a^*a)^2 + \ldots + \lambda_{n-1} (a^*a)^n.$$

Therefore we obtain

$$(p + 2|a^*a| 1_{\mathcal{V}})^{n+1} = \lambda_0 (p + 2|a^*a| 1_{\mathcal{V}}) + \ldots + \lambda_{n-1} (p + 2|a^*a| 1_{\mathcal{V}})^n$$

and this implies that $\mathcal{I}$ generates a finite dimensional invariant (for a^*a) subspace $\mathcal{Z}$ of $\mathcal{V}$. Since $\mathcal{I}_p(a^*a/_{\mathcal{Z}})$ is a finite subset of $\mathcal{I}_p(a^*a)$ and $\mathcal{Z} = \Sigma_\lambda \mathcal{V}_\lambda \cap \mathcal{Z}\ (\lambda \in \mathcal{I}_p(a^*a)/_{\mathcal{Z}}))$ we have the contradiction $y \in \Sigma_\lambda \mathcal{V}_\lambda\ (\lambda \in \mathcal{I}_p(a^*a))$. Thus iv) is proved.

Now let $\mathcal{B}_\lambda$ denote and arbitrary orthogonal basis of $\mathcal{V}_\lambda$, $\lambda \in \mathcal{I}_p(a^*a)$, and

$$\mathcal{B} = \cup_\lambda \mathcal{B}_\lambda \qquad (\lambda \in \mathcal{I}_p(a^*a)).$$

Whenever $x, y \in \mathcal{B}$ with $x \neq y$, we have

$$\langle a(x), a(y)\rangle = \langle x, a^*a(y)\rangle = k\langle x, y\rangle = 0$$

(y is supposed to be an element of $\mathcal{V}_k$). It follows that $a(\mathcal{B})$ is a subset of $\mathcal{W}$ whose elements are mutually orthogonal. Then there exists an orthogonal basis of $\mathcal{W}$, $\mathcal{R}$, such that

$$\mathcal{R} \supset \{\frac{u}{|u|} : u \in a(\mathcal{R}) - \{0\}\}.$$

<u>COROLLARY 1</u>. For any element a of the J*-algebra j, the norm $|a|$ is given by:

$$|a| = \sqrt{\rho(a^*a)}\ .$$

<u>COROLLARY 2</u>. Let a be any element of a J*-algebra. Then

$$|aa^*a| = |a|^3.$$

Any element a of a J*-algebra such that $aa^*a = a$ will henceforth be called <u>idempotent</u> (a is a partial isometry, see [2], p. 18).

The above definitions entails that if $a \neq 0$ is idempotent then $|a| = 1$.

<u>LEMMA 2</u>. The set of idempotent elements of a J*-algebra j generates j as a vector space. (There exists no proper subspace of j containing all idempotent elements of j).

<u>PROOF</u>. Let j be a non zero J*-algebra of bounded linear operators of the Hilbert space V into the Hilbert space W and let a be a non zero element of j. Whenever $\lambda \in I_p(a^*a) - \{0\}$, let $V_\lambda = \mathrm{Ker}(a^*a - \lambda 1_V)$ and let π_λ be the orthogonal projection of V onto V_λ. We prove that:

i) $I_p(a^*a)$ is a finite set;

ii) $|\lambda|^{-1/2} a \circ \pi_\lambda \in j$ whenever $\lambda \in I_p(a^*a) - \{0\}$;

iii) $|\lambda|^{-1/2} a \circ \pi_\lambda$ is idempotent whenever $\lambda \in I_p(a^*a) - \{0\}$;

iv) $a = \Sigma_\lambda\, a \circ \pi_\lambda \quad (\lambda \in I_p(a^*a) - \{0\})$.

Since $\{aa((aa^*)^n a)\} = (aa^*)^{n+1} a$, we have that $(aa^*)^m a \in j$ for $m = 0, \ldots,$ $((aa^*)^0 a = a)$. Assume $I_p(a^*a)$ has $n+1$ elements; we prove that the set

$$\{(aa^*)^m a : m = 0, 1, \ldots, n-1\} \subset j$$

is lineraly independent. Let $\mu_0, \ldots, \mu_{n-1}$ be n coefficient such that

$$0 = \mu_0 a + \ldots + \mu_{n-1}(aa^*)^{n-1} a$$

and let $\lambda \in I_p(a^*a) - \{0\}$ and $x \in V_\lambda - \{0\}$. Since

$$0 = (\mu_0 + \lambda\mu_1 + \ldots + \lambda^{n-1}\mu_{n-1})\, a(x),$$

then

$$(1) \quad 0 = \mu_0 + \lambda\mu_1 + \ldots + \lambda^{n-1}\mu_{n-1} \quad \text{whenever} \quad \lambda \in I_p(a^*a) - \{0\}.$$

The system (1) of n linear homogeneous equations, of the n unknowns $\mu_0,\ldots,\mu_{n-1}$, has a unique solution since the matrix of the coefficients of the unknowns is regular. It follows that

$$\text{Card}\, I_p(a^*a) \leqslant \dim j + 1$$

(Card $I_p(a^*a)$ denotes the number of members of $I_p(a^*a)$), and therefore $I_p(a^*a)$ is a finite set.

Let $\lambda \in I_p(a^*a) - \{0\}$ and $x \in V_\lambda - \{0\}$; we have

$$\lambda < x,x > = < a^*a(x),x > = < a(x),a(x) > > 0.$$

Hence $I_p(a^*a)$ is a subset of $\mathbb{R}^+$. Let now ρ denote the spectral radius of a^*a (ρ is strictly positive) and let us consider the sequence of elements of j defined inductively by.

$$b_1 = \rho^{-1/2}\, a,$$

$$b_{n+1} = b_n b_n^* b_n = \rho^{-1/2}\, a(\rho^{-1}\, a^*a)^{(3^n-1)/2}.$$

For $x \in V$, we have

$$b_{n+1}(x) = \Sigma_\lambda (\lambda/\rho)^{(3^n-1)/2}\, \rho^{-1/2}\, a\, \pi_\lambda(x) \quad (\lambda \in I_p(a^*a))$$

and hence, if $|x| = 1$ and $n > 1$,

$$|b_{n+1}(x) - b_n(x)| \leqslant \Sigma_\lambda |(\lambda/\rho)^{(3^n-1)/2} - (\lambda/\rho)^{(3^{n-1}-1)/2}|\rho^{-1/2}|a| \leqslant$$

$$\leqslant \Sigma_\mu 2(\mu/\rho)^{(3^{n-1}-1)/2} \leqslant 2(\nu/\rho)^{(3^{n-1}-1)/2} \text{ Card } I_p(a^*a) \leqslant$$

$$\leqslant 2(\nu/\rho)^{3^{n-2}} \text{ Card } I_p(a^*a) \quad (\lambda \in I_p(a^*a),\ \mu \in I_p(a^*a) - \{\rho\})$$

(where $\nu = \sup(I_p(a^*a) - \{\rho\}) \cup \{0\})$. It follows that

$$|b_{n+h} - b_n| \leqslant 2 \text{ Card } I_p(a^*a)\ (\nu/\rho)^{3^{n-2}} \ (1 + (\nu/\rho)^3 + \dots (\nu/\rho)^{3h})$$

$$\leqslant 2 \text{ Card } I_p(a^*a)\ (\nu/\rho)^{3^{n-2}} \ (\rho^3/(\rho^3 - \nu^3))$$

hence $b_1, b_2, \dots, b_n, \dots$ is a Cauchy sequence. Since

$$j \ni \lim_{n\to\infty} b_n = \lim_{n\to\infty} \Sigma_\lambda (\lambda/\rho)^{(3^n-1)/2} \rho^{-1/2} a\pi_\lambda = \rho^{-1/2} a\pi_\rho,$$

ii) is proved provided that Card $I_p(a^*a) = 1$.
Now assume Card $I_p(a^*a) > 1$, and define $a_1 = a - a\pi_\rho \in j$; we have

(2) $I_p(a_1^*a_1) - \{0\} = I_p(a^*a) - \{0,\rho\}$;

(3) $\text{Ker}(a_1^*a_1 - \lambda 1_V) = \text{Ker}(a^*a - \lambda 1_V) = V_\lambda$, whenever $\lambda \in I_p(a_1^*a_1) - \{0\}$;

(4) $a_1\pi_\lambda = a\pi_\lambda$ whenever $\lambda \in I_p(a_1^*a_1) - \{0\}$.

Arguing as above, we can prove that $a\pi_\lambda \in j$ where $\nu = \sup(I_p(a^*a) - \{\rho\})$.
A finite inductive argument produces the proof of ii).

iii) and iv) are trivial consequences of ii).

An idempotent element of a J*-algebra j will be called <u>irreducible</u> if

$$3\{aa\{aab\}\} - \{aab\} = aa^*ba^*a + ab^*a \doteq a, \text{ whenever } b \in j$$

($\doteq$ means proportional to).

Let a be any idempotent element of J*-algebra j. The mapping $\psi_a : j \to j$ defined by

$$\psi_a(x) = 4\{aax\} - 3\{aa\{aax\}\} = aa^*x + xa^*a - aa^*xa^*a$$

is a projection.

Two idempotent elements a, b of a J*-algebra are said to be <u>strongly independent</u> if $\psi_a(b) = 0$.

<u>LEMMA 3</u>. Let a, b be two idempotent elements of a J*-algebra j. The following properties hold.

A) if $\psi_a(b) = 0$ then $a^*b = b^*a = 0$ and $ab^* = ba^* = 0$;

B) if $\psi_a(b) = 0$ then $\psi_b(a) = 0$;

C) if a and b are strongly independent then $\psi_a\psi_b = \psi_b\psi_a$;

D) if a and b are non zero, irreducible and such that $ab^*a = 0$ and $\psi_a(b) = b$ then $ba^*a = 0$ and $\psi_b(a) = a$.

<u>PROOF</u>. A) follows from

$$0 = \psi_a(b) = aa^*b + ba^*a - aa^*ba^*a$$

by multiplying both sides of this equation, first on the left and then on the right, by a*. B) is a trivial consequence of A).
If $c \in j$, A) yields

$$\psi_a\psi_b(c) = aa^*(bb^*c + cb^*b - bb^*cb^*b) + (bb^*c + cb^*b - bb^*cb^*b)a^*a$$

$$- aa^*(bb^*c + cb^*b - bb^*cb^*b)a^*a = aa^*cb^*b + bb^*ca^*a.$$

Replacing a by b and b by a we obtain

$$\psi_b\psi_a(c) = aa^*cb^*b + bb^*ca^*a,$$

whence C) follows.

The assumption in D) yields $b = aa^*b + ba^*a$ and $bb^*ab^*b + ba^*b = b$. Hence $a^*bb^*ab^*ba + a^*ba^*ba = a^*ba^*, a^*bb^*ab^*ba^* = 0$, and therefore

$$0 = aa^*bb^*ab^*ba^*a = (b - ba^*a)b^*ab^*(b - aa^*b) = bb^*ab^*b.$$

Then we have $ba^*b = 0$ and $\psi_b(a) = bb^*a + ab^*b$. Moreover, since a is irreducible, it is

$$aa^*(bb^*a + ab^*b)a^*a + a(b^*ba^* + a^*bb^*)a = \lambda a$$

and

$$2(aa^*bb^*a + ab^*ba^*a) = 2(bb^*a + ab^*b) = \lambda a.$$

Multiplying the last relation on the left by b^* we obtain $2ab^*bb^* = \lambda ab^*$ which implies $\lambda = 2$ and $a = bb^*a + ab^*b = \psi_b(a)$.

<u>LEMMA 4</u>. Let $\mathcal{B}$ be a subset of a J*-algebra $\mathcal{J}$ whose elements are non zero, idempotent and mutually strongly independent. $\mathcal{B}$ is a linearly independent subset of $\mathcal{J}$.

<u>PROOF</u>. Let $a_1, \dots, a_n$ be n distinct elements of $\mathcal{B}$ and assume that $\mu_1 a_1 + \dots + \mu_n a_n = 0$. For $i = 1, \dots, n$, we have

$$0 = (\mu_1 a_1 + \dots + \mu_n a_n)a_i^* a_i = \mu_i a_i$$

and hence $\mu_i = 0$.

<u>LEMMA 5</u>. The set of all idempotent irreducible elements of a J*-algebra $\mathcal{J}$ generates $\mathcal{J}$, as a vector space.

<u>PROOF</u>. Let a be a non zero element of $\mathcal{J}$. With the same notation used in the proof of lemma 2, we have

$$a = \Sigma_\lambda \sqrt{\lambda}\frac{a}{\sqrt{\lambda}}\pi_\lambda \qquad (\lambda \in I_p(a^*a) - \{0\}).$$

Hence a is a linear combination of $n = \mathrm{Card}(I_p(a^*a) - \{0\})$ non zero idempotent, mutually strongly independent elements of j. Let us consider the subset B of j whose elements are non zero, idempotent, mutually strongly independent and such that $a \in V(B)$ ($V(B)$ is the minimum subspace of j containing B).

All these sets are finite sets with a number of elements not greater than the dimension of j. Let B be one of the above sets with a maximal number of elements. We will prove that any element of B is irreducible.

Assume B has an element b that is not irreducible and let c be an element of j such that

$$\vartheta_b(c) = bb^*cb^*b + bc^*b$$

is not proportional to b. Since b is assumed to be idempotent, we have

$$(1)\quad \vartheta_b(c)^*b = b^*bc^*b + b^*cb^*b = b^*\vartheta_b(c) = (\vartheta_b(c)^*b)^*.$$

Now, proceeding as in the proof of lemma 2, we obtain:

i) $I_p(\vartheta_b(c)^*b) \subset \mathbb{R}$;

and, setting $V_\lambda = \mathrm{Ker}(\vartheta_b(c)^*b - \lambda 1_V)$,

ii) $V_\lambda = \Sigma_\lambda V_\lambda \quad (\lambda \in I_p(\vartheta_b(c)^*b))$

(V is the Hilbert space where the elements of j are defined)

iii) $V_\lambda \perp V_\mu$, whenever $\lambda, \mu \in I_p(\vartheta_b(c)^*b)$ with $\lambda \neq \mu$.

Now let P_λ be any orthonormal basis of V_λ; then $P = \cup_\lambda P_\lambda \quad (\lambda \in I_p(\vartheta_b(c)^*b))$ is an orthonormal basis of V. Define $Z = \mathrm{Ker}(b^*b - 1)$ and let π_Z be the orthonormal projection of V onto Z. Lemma 2 implies that $bb^* = \pi_Z$ and, since $b^*b\vartheta_b(c)^*bb^*b = \vartheta_b(c)^*b$,

$$\cup_\lambda P_\lambda \subset Z \qquad (\lambda \in I_p(\vartheta_b(c)^*b)).$$

The last relation entails the existence of an orthonormal basis of W (the Hilbert space where the elements of j take their values), R, such that

$$R \supset \{ \frac{u}{|u|} : u \in b(P) - \{0\}\}.$$

Moreover, whenever $v \in P$ we have $\vartheta_b(c)(v) \doteq b(v)$. Set now $\mu = \max\{|\vartheta_b(c)(v)| : v \in P\}$. Then $\mu > 0$. Let υ be a scalar such that $|\upsilon| = \mu$. There exists an element v in P such that $\vartheta_b(c)(v) = \upsilon b(v)$. Since $\vartheta_b(c)$ is not proportional to b, the space

$$Y = \{x : x \in Z \text{ and } \vartheta_b(c)(x) = \upsilon\, b(x)\}$$

is a proper subspace of Z. Consider now $h = 1/2\ (b + \upsilon^{-1}\vartheta_b(c))$ and the two idempotent elements of j

$$\lim_{n\to\infty} (hh^*)^n h = b\pi_Y$$

$$b - b\pi_Y = b\pi_{Y^\perp}$$

It is now trivial to prove that

$$Q = (B - \{b\}) \cup \{b\pi_Y, b\pi_{Y^\perp}\}$$

is a set of non zero, idempotent, mutually strongly independent elements of j such that $V(Q) \ni a$ and $\text{Card}\, Q = \text{Card}\, B + 1$. We have thus obtained a contradiction. Hence any element of B is irreducible.

LEMMA 6. Let a and b be two non zero, idempotent, strongly independent elements of a J*-algebra j. Then:

A) $\text{Ker}\, a^\perp \perp \text{Ker}\, b^\perp$;

B) $\text{Im}\, a \perp \text{Im}\, b$.

PROOF. Let $\mathcal{J}$ be a J*-algebra of bounded linear operators of the Hilbert space V into the Hilbert space W. Whenever $x,y \in V$ then

$$0 = \langle a^*b(x),y\rangle = \langle a(x),b(y)\rangle$$

and this proves B). A) follows from B) and from Ker a = Im a, Ker b = Im b.

LEMMA 7. Any J*-isomorphism between two J*-algebras is an isometry.

PROOF. It is a trivial consequence of the previuos lemma since any J*-isomorphism preserves strong independence.

LEMMA 8. Let a and b be two nonzero, idempotent, strongly indepen dent elements of a J*-algebra $\mathcal{J}$. The set

$$\psi_a\psi_b(\mathcal{J}) = \{aa^*xb^*b + bb^*xa^*a : x \in \mathcal{J}\}$$

is a J*-sub algebra (a subset of $\mathcal{J}$ that is a J*-algebra) of $\mathcal{J}$.

LEMMA 9. Let a and b be two non zero, idempotent, irreducible and strongly independent elements of a J*-algebra $\mathcal{J}$ such that $\psi_a\psi_b(\mathcal{J}) \neq \{0\}$. Then there exists a non zero idempotent element $c \in \psi_a\psi_b(\mathcal{J})$ such that

A) $6\{abc\} = \epsilon c$ where $\epsilon \in \{+1, -1\}$.

Furthermore, any non zero idempotent element $c \in \psi_a\psi_b(\mathcal{J})$ satisfying A) satisfies also

B) $cc^*a = ac^*c = a$;

C) $cc^*b = bc^*c = b$;

D) $ca^*c = \epsilon b$;

E) $cb^*c = \epsilon a$.

PROOF. By lemma 2, $\psi_a\psi_b(j)$ has at least one non zero idempotent element, k.

If $6\{abk\} + k = 0$, then $c = k$ and $\epsilon = -1$ satisfies A).

Assume now $s = 6\{abk\} - k \neq 0$; then

(1) $6\{abs\} = s$.

The four elements of j:

$$\psi_a(3\{ass\}) = as^*s + ss^*a;$$

$$\psi_b(3\{ass\}) = sa^*a;$$

$$\psi_b(3\{bss\}) = bs^*s + ss^*b;$$

$$\psi_a(3\{bss\}) = sb^*s,$$

together with (1) and the irreducibility of a and b, give

$$(2)\begin{cases} as^*s \doteq a, & ss^*a \doteq a, \\ bs^*s \doteq b, & ss^*b \doteq b, \\ sa^*s \doteq b, & sb^*s \doteq a. \end{cases}$$

Let $as^*s = \lambda a$ and $bs^*s = \mu b$. We have

$$\lambda a = as^*bb^*s = sb^*s = sb^*sa^*a = ss^*a$$

and

$$\mu b = sa^*s = ss^*b.$$

It follows that $\lambda as^* = sb^*ss^* = \mu sb^*$ which implies $\lambda^2 = \mu\lambda$. Now

$$ss^*s = sa^*sb^*s + sb^*sa^*s = \lambda(sa^*a + aa^*s) = \lambda s$$

and lemma 1 yield $\lambda = \mu > 0$. $c = \lambda^{-1/2} s$ and $\epsilon = 1$ satisfy A).

Assume now c is a nonzero, idempotent element of $\psi_a \psi_b(j)$ which satisfies A). c satisfies also relations analogous to (2). Let now $ac^*c = \lambda a$. We have $ac^*cc^* = \lambda ac^* = ac$ and then $ac^*c = a$. The equalities B) ... E) can be established in a similar way.

LEMMA 10. Let a be any non zero, irreducible idempotent element of a J*-algebra j. If j has a non zero, irreducible, idempotent element b such that a and b are strongly independent and $\psi_a \psi_b(j) \neq \{0\}$, then:

A) $aa^*xa^*a \in j$ whenever $x \in j$;

B) the mapping δ_a defined by $\delta_a(x) = aa^*xa^*a$ is a projection on j;

C) $\delta_a(j)$ is a J*-sub algebra of j containing a.

PROOF. Let $c \in \psi_a \psi_b(j)$ be a non zero, idempotent element such that $6\{abc\} = \epsilon c$, with $\epsilon \in \{1, -1\}$. Let $\tau_c : j \to j$ be the linear mapping defined by $\tau_c(x) = 3\{ccx\}$. Then, for any $x \in j$, we have that $\psi_a \tau_c (\psi_a - \psi_a \psi_b) \tau_c \psi_a(x) = aa^*xa^*a = \delta_a(x)$ is an element of j and A) is proved. B) and C) are now trivially checked.

LEMMA 11. Let j be a J*-algebra. If j has a non zero, idempotent, irreducible element a such that δ_a is the identity on j, then we have:

A) whenever $x \in j$ then $xx^* = \lambda aa^*$, $x^*x = \lambda a^*a$;

B) whenever $x \in j$ then $xx^*x = \lambda x$;

C) whenever x is a nonzero, idempotent element of j then δ_x is the identity on j;

D) any idempotent element of j is irreducible;

E) j has a basis $\{a_0, \ldots, a_{n-1}\}$ whose elements are idempotent and satisfy $a_i a_j^* + a_j a_i^* = 0$, $a_i^* a_j + a_j^* a_i = 0$ whenever $i, j = 1, \ldots, n-1$ and $i \neq j$.

F) for all $x,y \in j$, $xy^*x \in j$;

G) for all $x,y \in j - \{0\}$, there exists $z \in j$ such that $zx^*z = y$.

PROOF. Since a is irreducible, we have $aa^*xa^*a + ax^*a = x + ax^*a \doteq a$ hence $3\{axx\} - xx^*a \doteq x$. Then $xx^*a \in \mathcal{J}$, $xx^*a + aa^*xx^*a = 2xx^*a \doteq a$ and therefore $xx^* \doteq aa^*$. If $xx^* = \lambda aa^*$, then $xx^*x = \lambda aa^*x = \lambda x$ and B) is proved. A) and C) follow easily. We define now inductively a sequence of J*-algebras and of elements of $\mathcal{J}$, as follows:

$$\mathcal{J}_0 = \mathcal{J} \quad \text{and} \quad a_0 = a,$$

$$\mathcal{J}_{n+1} = \{b : b \in \mathcal{J}_n \text{ such that } b + a_n b^* a_n = 0\} \quad \text{and}$$

$$a_{n+1} = 0 \quad \text{if} \quad \mathcal{J}_{n+1} = \{0\},$$

$$a_{n+1} \text{ is an idempotent, non zero element of } \mathcal{J}_{n+1} \text{ irreducible}$$

$$\text{in } \mathcal{J}_{n+1}, \text{ if } \mathcal{J}_{n+1} \neq \{0\}.$$

Let p and q be two integers such that $0 \leq p < q$, we have $\mathcal{J}_p \subset \mathcal{J}_q$ and therefore $a_q + a_p a_q^* a_p = 0$. The last equality together with A) yields

$$a_p a_q^* + a_q a_p^* = 0 \quad \text{and} \quad a_p^* a_q + a_q^* a_p = 0.$$

Now we prove, by an inductive argument, that

$$(1) \quad \mathcal{J} = V(\{a_0, \ldots, a_{m-1}\}) + \mathcal{J}_m \quad \text{whenever} \quad m \geq 0.$$

Clearly (1) holds if $m = 0$. Assume it holds when $m = r$. Since a_r is an idempotent irreducible element of $\mathcal{J}_r$, for all $b \in \mathcal{J}_r$, we have $b + a_r b^* a_r = a_r$ and therefore

$$(b - \lambda/2\, a_r) + a_r(b - \lambda/2\, a_r)^* a_r = 0.$$

Thus $b - \lambda/2\, a_r \in \mathcal{J}_{r+1}$. It follows that $b \in V(\{a_r\}) + \mathcal{J}_{r+1}$. Thus (1) is proved. Assume $a_0, \ldots, a_m \neq 0$. If $\lambda_s a_s + \ldots + \lambda_m a_m = 0$ for some $0 \leq s \leq m$ and some scalars $\lambda_s, \ldots, \lambda_m$, then

$$0 = \bar{\lambda}_s(\lambda_s a_s + \dots + \lambda_m a_m) + \lambda_s a_s(\lambda_s a_s + \dots + \lambda_m a_m)^* a_s = 2|\lambda_s|^2 a_s$$

and $\lambda_s = 0$. This proves that $a_0,\dots,a_m$ are linearly independent. Let $n-1$ be the first integer such that $j_n = \{0\}$. $a_0,\dots,a_{n-1}$ is a basis that satisfies E).

For any $x = \lambda_0 a_0 + \dots + \lambda_{n-1} a_{n-1}$ and $i = 0,\dots,n-1$, we have

$$x a_i^* x = \Sigma_{r,s}\, \lambda_r \lambda_s a_r a_i^* a_s = -\Sigma_{p,s}\, \lambda_p \lambda_s a_i a_p^* a_s + \Sigma_s\, \lambda_i \lambda_s a_s = -a_i + 2\lambda_i x$$

$$(p,r,s = 0,\dots,n-1 \text{ and } p \neq i).$$

Hence,

$$xx^* a_i x^* x + x a_i^* x = -x a_i^* x + 2\lambda_i x + x a_i^* x = 2\lambda_i x,$$

and D) follows by linearity. F) is a consequence of C) and D). Let $x = \Sigma_i \lambda_i a_i$ $(i = 0,\dots,n-1)$, $y = \Sigma_i \mu_i a_i$ $(i = 0,\dots,n-1)$ and $z = \Sigma_i \tau_i a_i$ $(i = 0,\dots,n-1)$. Condition

(2) $zx^*z = \epsilon y$ with $\epsilon \in \{1, -1\}$

holds if and only if $x + \epsilon y = 2(\Sigma_i \tau_i \lambda_i) z$ $(i = 0,\dots,n-1)$. If we set $\alpha = 2\Sigma_i \tau_i \lambda_i$ $(i = 0,\dots,n-1)$ (2) holds if and only if, the system

$$\lambda_j + \epsilon\mu_j = \alpha\tau_j \quad j = 0,\dots,n-1$$

has a solution. Let $\epsilon \in \{1, -1\}$ be chosen in such a way that $2\Sigma_i(\lambda_i + \epsilon\mu_i)\lambda_i > 0$ $(i = 0,\dots,n-1)$. (Since $y \neq 0$, this is always possible). Choosing $\alpha = (2\Sigma_i(\lambda_i + \epsilon\mu_i)\lambda_i)^{1/2}$ $(i = 0,\dots,n-1)$, it is easily checked that

$$z = \alpha^{-1}\Sigma_i(\lambda_i + \epsilon\mu_i)\lambda_i a_i \quad (i = 0,\dots,n-1)$$

satisfies G).

LEMMA 12. Let a and b be two non zero, irreducible, strongly idempotent elements of a J*-algebra j such that the J*-algebra $\psi_a\psi_b(j)$ is not zero. The following propositions are equivalent:

A) $\psi_a\psi_b(j)$ has a non zero, idempotent element c which is irreducible in $\psi_a\psi_b(j)$ and such that the restriction to $\psi_a\psi_b(j)$ of δ_c is the identity;

B) $\psi_a\psi_b(j)$ has a non zero idempotent element c, irreducible in $\psi_a\psi_b(j)$, such that $6\{abc\} = \epsilon c$ with $\epsilon \in \{1, -1\}$;

C) any non zero, idempotent, irreducible element of j is not contained in $\psi_a\psi_b(j)$.

PROOF. A) ⇒ B). Lemma 9 implies the existence of a non zero, idempotent element $c \in \psi_a\psi_b(j)$ such that $6\{abc\} = \epsilon c$ with $\epsilon \in \{1, -1\}$. c is irreducible, in $\psi_a\psi_b(j)$, by D) of lemma 11.

B) ⇒ A). Let $x \in \psi_a\psi_b(j)$. We have $cc^*xc^*c = cc^*(aa^*xb^*b + bb^*xa^*a)c^*c$. The above identity and lemma 8 give $cc^*xc^*c = \delta_c(x) = x$.

A) ⇒ C). Let x be a non zero, idempotent element of $\psi_a\psi_b(j)$. We have (lemma 11) $\psi_a(xx^*ax^*x + xa^*x) = \dot{\psi}_a(a + xa^*x) = a$. Since $a \neq 0$ and a is not in $\psi_a\psi_b(j)$, x and $xx^*ax^*x + xa^*x$ cannot be proportional. Then x is irreducible in j.

non B) ⇒ non C). Let c be a non zero, idempotent element of $\psi_a\psi_b(j)$ which is irreducible in $\psi_a\psi_b(j)$. Let us prove that is

$$cf^*c = cg^*c = 0 \quad \text{whenever} \quad f \in \delta_a(j) \quad \text{and} \quad g \in \delta_b(j).$$

Assume $cf^*c \neq 0$; then there exist an element $h \in \delta_a(j)$ and an element $t \in \delta_b(j)$ such that $ha^*h = \epsilon f$ and $tc^*fc^*t = \eta b$ where $\epsilon, \eta \in \{1, -1\}$ (See lemma 11, G)). Let k be the element of $\psi_a\psi_b(j)$ defined by $k = 6\{htc\} = hc^*t + tc^*h$. Then

$$(1) \quad ka^*k = \epsilon\eta b \quad \text{and} \quad 6\{abk\} = \epsilon\eta k.$$

For any $p \in \psi_a\psi_b(j)$, we have

$$kk^*pk^*k + kp^*k = hc^*tt^*ch^*pt^*ch^*hc^*t + tc^*hh^*ct^*ph^*ct^*tc^*h +$$

$$+ hc^*tp^*hc^*t + tc^*hp^*tc^*h = 6h(c^*c\{thp\}^*cc^* + c^*\{thp\}c^*)t +$$

$$+ 6t(c^*c\{thp\}^*cc^* + c^*\{thp\}c^*)h \doteq hc^*t + tc^*h = k.$$

The above relation together with (1) contradicts the assumption not B) and hence $cf^*c = 0$. Similarly we can prove $cg^*c = 0$.
For $x \in j$, we have

$$cc^*xc^*c + cx^*c = cc^*aa^*xa^*ac^*c + cc^*bb^*xb^*bc^*c +$$

$$+ cc^*(aa^*xb^*b + bb^*xa^*a)c^*c + ca^*ax^*aa^*c + cb^*bx^*bb^*c +$$

$$+ c(a^*ax^*bb^* + b^*bx^*aa^*)c = cc^*\delta_a(x)c^*c + c\delta_a(x)^*c +$$

$$+ cc^*\delta_b(x)c^*c + c\delta_b(x)^*c + cc^*\psi_a\psi_b(x)c^*c + c\psi_a\psi_b(x)^*c \doteq c .$$

<u>LEMMA 13</u>. Let a and b be two non zero, idempotent irreducible elements of a J*-algebra j which are strongly independent and $\psi_a\psi_b(j) \neq \{0\}$.
The following statements are equivalent:

A) $\psi_a\psi_b(j)$ has a non zero idempotent element which is irreducible in j;

B) any non zero idempotent element of $\psi_a\psi_b(j)$ which is irreducible in $\psi_a\psi_b(j)$ is irreducible in j;

C) $\psi_a\psi_b(j)$ has two non zero idempotent elements which are irreducible in $\psi_a\psi_b(j)$ and strongly independent;

D) $\psi_a\psi_b(j)$ has a non zero idempotent element c, irreducible in $\psi_a\psi_b(j)$ and such that $ca^*c = cb^*c = 0$;

E) if c is a non zero, idempotent, element of $\psi_a\psi_b(j)$ which is irreducible in $\psi_a\psi_b(j)$ then $6\{abc\}$ is a non zero, idempotent, irreducible element if $\psi_a\psi_b(j)$, moreover c and $6\{abc\}$ are strongly independent;

F) whenever c is a non zero idempotent irreducible element of

$\psi_a\psi_b'(j)$ then $ca^*c = cb^*c = 0$.

PROOF. The implications B) ⇒ A), F) ⇒ D) and E) ⇒ C) are consequences of lemma 2. Our condition A) is condition non B) of lemma 12. Hence the proof of lemma 12 gives us A) ⇒ F) ⇒ B) and D) ⇒ A). If condition C) holds, there is no non zero element c of $\psi_a\psi_b(j)$ which is irreducible in $\psi_a\psi_b(j)$ and such that $6\{abc\} = \epsilon c$ with $\epsilon \in \{1, -1\}$ (lemma 12). Hence C) ⇒ A). Finally assume that F) holds and let $c \in \psi_a\psi_b(j)$ be a non zero idempotent element irreducible in $\psi_a\psi_b(j)$ such that $ca^*c = cb^*c = 0$. Then $6c^*\{abc\} = c^*(ac^*b + bc^*a) = 0$ and $6\{abc\}c^* = (ac^*b + bc^*a)c^* = 0$. Hence c and $6\{abc\}$ are strongly independent. It is easy to prove that $6\{abc\}$ is a non zero idempotent element of $\psi_a\psi_b(j)$ irreducible in $\psi_a\psi_b(j)$. F) ⇒ E) is then proved.

COROLLARY 3. Let a and b be two non zero idempotent irreducible and strongly independent elements of a J*-algebra j such that $\psi_a\psi_b(j) \neq \{0\}$. If $\psi_a\psi_b(j)$ has a non zero idempotent elements c, which is irreducible in j, then

A) $cc^*a + ac^*c = a$;

B) $cc^*b + bc^*c = b$.

Whenever a is a non zero idempotent element of a J*-algebra j, we shall denote by γ_a the projection of j defined by $\gamma_a = 1_j - \psi_a$.

It is easily checked that:

LEMMA 14. Let a be a non zero idempotent element of a J*-algebra j. Then $\gamma_a(j)$ is a J*-subalgebra of j.

A J*-algebra j is called reducible if j has two non zero, idempotent, irreducible, strongly independent elements a and b such that $\psi_a\psi_b(j) = \{0\}$.
Otherwise j is said to be irreducible.

LEMMA 15. Let a, b and c be three non zero, idempotent, irreducible, pairwise strongly independent elements of a J*-algebra j . If $\psi_a\psi_b(j) \neq \{0\}$ and $\psi_b\psi_c(j) \neq \{0\}$ then $\psi_a\psi_c(j) \neq \{0\}$.

PROOF. Lemma 9 implies the existence of two non zero, idempotent elements $h \in \psi_a\psi_b(j)$, $k \in \psi_b\psi_c(j)$ such that $6\{abh\} = \epsilon h$ and $6\{bck\} = \eta k$ where $\epsilon, \eta \in \{1, -1\}$. Let $t = \{hkb\}$. We have

$$\psi_a\psi_c(t) = aa*(hb*k + kb*h)c*c + cc*(hb*k + kb*h)a*a = t.$$

Then, since

$$6\{thc\} = ck*bh*h + hh*bk*c = ck*b + bk*c = \eta k \neq 0$$

we have $t \neq 0$.

LEMMA 16. If j is a J*-algebra, there exists a finite set of irreducible J*-sub algebras of j , $\{j_1,\ldots,j_n\}$ such that

$$j = j_1 \oplus \cdots \oplus j_n.$$

This decomposition is unique.

PROOF. Let A be a subset of j such that:

i) $0 \notin A$;

ii)   any element of A is idempotent;

iii) any element of A is irreducible;

iv)   for all $x, y \in A$, $\psi_x\psi_y(j) \neq \{0\}$;

v)   if $x, y \in A$, $x \neq y$, then x and y are strongly independent;

vi)   A is a maximal set (in the partial ordering given by inclusion).

Lemma 15, in view of vi), entails that any non zero, idempotent element of j which is strongly independent of any element of A is strongly independent from any non zero, idempotent element of $\psi_y(j)$ ($y = \Sigma_x \, x \; (x \in A)$).

Let us denote by EA the set of non zero, idempotent irreducible elements of $\mathcal{J}$ which are strongly independent of any element of A. The two J*-algebras:

$$\mathcal{J}_A = \psi_t(\mathcal{J}) = \cap_y \gamma_y(\mathcal{J}) \quad (t = \Sigma_x\, x\ (x \in A) \text{ and } y \in A);$$

$$\mathcal{J}_{EA} = V(EA) = \gamma_t(\mathcal{J}) = \cap_y \gamma_y(\mathcal{J}) \quad (t = \Sigma_x\, x\ (x \in A) \text{ and } y \in A)$$

are two complementary subspaces of $\mathcal{J}$ whose direct sum is $\mathcal{J}$. Moreover any element of $\mathcal{J}_A$ is strongly independent of any element of $\mathcal{J}_{EA}$. Hence the set $\mathcal{J}_A \cup \mathcal{J}_{EA}$ contains every irreducible element of $\mathcal{J}$. Let now A' be a new set which satisfies i), ..., vi), two cases are possible: $\mathcal{J}_A = \mathcal{J}_{A'}$ or $\mathcal{J}_{A'} \subset \mathcal{J}_{EA}$. Therefore we either have $\mathcal{J}_A = \mathcal{J}_{A'}$ or $\mathcal{J}_A \cap \mathcal{J}_{A'} = \{0\}$. Hence there exists a finite set $\{A_1, \ldots, A_n\}$ of subsets of $\mathcal{J}$ such that:

A_i satisfies i), ..., vi), whenever $i = 1, \ldots, n$;

$\mathcal{J}_{A_i} \cap \mathcal{J}_{A_h} = \{0\}$, for all $1 \leqslant i < h \leqslant n$;

$\mathcal{J}_{A_i}$ is irreducible whenever $i = 1, \ldots, n$;

$\mathcal{J} = \mathcal{J}_{A_1} \oplus \ldots \oplus \mathcal{J}_{A_n}$.

Let V_1 and V_2 be two Hilbert vector space on $\mathbb{K}$. Henceforth $V_1 \oplus V_2$ will denote the Hilbert vector space on $\mathbb{K}$ given by the direct sum of V_1 and V_2 endowed with the norm

$$|x + y| = (|x|^2 + |y|^2)^{1/2}, \quad \text{for } x \in V_1 \text{ and } y \in V_2.$$

Let $\mathcal{J}_i$ be a J*-algebra of bounded linear operator of V_i into W_i $(i = 1,2)$. Moreover let π_{W_i} be the orthogonal projection of $W_1 \oplus W_2$ on W_i. The set of bounded linear operators of $V_1 \oplus V_2$ into $W_1 \oplus W_2$ given by

$$\mathcal{j}_1 \oplus \mathcal{j}_2 = \{a : a \in L(V_1 \oplus V_2, W_1 \oplus W_2) \text{ such that } \pi_{W_1} a_{/V_1} \in \mathcal{j}_1 \text{ and } \pi_{W_2} a_{/V_2} \in \mathcal{j}_2\}$$

is a J*-algebra which we shall call the direct sum of $\mathcal{j}_1$ and $\mathcal{j}_2$. Since the additive structure of $\mathcal{j}_1 + \mathcal{j}_2$ is naturally identified with the direct sum of the additive structures of $\mathcal{j}_1$ and $\mathcal{j}_2$, then $\mathcal{j}_1$ and $\mathcal{j}_2$ are identified naturally with two subalgebras of $\mathcal{j}_1 \oplus \mathcal{j}_2$.

LEMMA 17. Let $\mathcal{j}$ be a J*-algebra. We have

A) $\mathcal{j}$ is reducible if, and only if, $\mathcal{j}$ is J*-isomorphic to the direct sum of two non zero J*-algebras;

B) if $\mathcal{j}_1, \ldots, \mathcal{j}_n$ are n distinct non zero irreducible J*-sub algebras of $\mathcal{j}$ such that $\mathcal{j} = \mathcal{j}_1 + \ldots + \mathcal{j}_n$, then $\mathcal{j}$ is J*-isomorphic to $\mathcal{j}_1 \oplus \ldots \oplus \mathcal{j}_n$.

Now we have to distinguish between $\mathbb{R}$-J*-algebras and $\mathbb{C}$-J*-algebras.

Let $\mathcal{j}$ be a $\mathbb{C}$-J*-algebra and a, b, c three elements of $\mathcal{j}$. It is easily checked that the element [abc] defined by

$$[abc] = \frac{1}{2}(ab^*c + cb^*a)$$

is in $\mathcal{j}$.

Let $\mathcal{j}$ be an $\mathbb{R}$-J*-algebra and $\mathcal{B}$ an $\mathbb{R}$-basis of $\mathcal{j}$. The mapping $\mu : \mathcal{B} \times \mathcal{B} \times \mathcal{B} \to \mathbb{R}$ defined by the equalities

$$(1) \quad [abc] = \Sigma_x \lambda(a,b,c,x)x \quad (x \in \mathcal{B}),$$

is called the (real) structure coefficient of $\mathcal{B}$. The (1) are called the (real) structure formulas of $\mathcal{B}$.

If $\mathcal{j}$ is a $\mathbb{C}$-J*-algebra and $\mathcal{B}$ a $\mathbb{C}$-basis of $\mathcal{j}$, the mappings $\mu : \mathcal{B} \times \mathcal{B} \times \mathcal{B} \to \mathbb{C}$ defined by the equalities

$$(2) \quad [abc] = \Sigma_x \mu(a,b,c,x)x \quad (x \in \mathcal{B})$$

are called the (<u>complex</u>) <u>structure coefficient</u> of $\mathcal{B}$ and the equalities (2) the (<u>complex</u>) <u>structure formulas</u> of $\mathcal{B}$.

Let j be a $\mathbb{C}$-J*-algebra and $\mathcal{B}$ a $\mathbb{C}$-basis of j. The subset of j

$$\mathcal{B} \cup i\mathcal{B} = \{u, iu : u \in \mathcal{B}\}$$

$(i = \sqrt{-1})$ is a $\mathbb{R}$-basis of j (considered as a $\mathbb{R}$-J*-algebra). The real structure coefficient of $\mathcal{B} \cup i\mathcal{B}$ and the complex structure coefficient of $\mathcal{B}$ are linked by the relations:

$$\lambda(a,b,c,x) = \frac{1}{3}\,\mathrm{Re}(\mu(a,b,c,x) + \mu(b,c,a,x) + \mu(c,a,b,x))$$

$$\lambda(a,b,c,ix) = \frac{1}{3}\,\mathrm{Im}(\mu(a,b,c,x) + \mu(b,c,a,x) + \mu(c,a,b,x))$$

$$\lambda(ia,b,c,x) = \frac{1}{3}\,\mathrm{Im}(-\mu(a,b,c,x) - \mu(b,c,a,x) + \mu(c,a,b,x))$$

$$\lambda(ia,b,c,ix) = \frac{1}{3}\,\mathrm{Re}(\mu(a,b,c,x) + \mu(b,c,a,x) - \mu(c,a,b,x))$$

$$\lambda(ia,ib,c,x) = \frac{1}{3}\,\mathrm{Re}(\mu(a,b,c,x) - \mu(b,c,a,x) + \mu(c,b,a,x))$$

$$\lambda(ia,ib,c,ix) = \frac{1}{3}\,\mathrm{Im}(\mu(a,b,c,x) - \mu(b,c,a,x) + \mu(c,b,a,x))$$

$$\lambda(ia,ib,ic,x) = \frac{1}{3}\,\mathrm{Im}(-\mu(a,b,c,x) - \mu(b,c,a,x) - \mu(c,b,a,x))$$

$$\lambda(ia,ib,ic,ix) = \frac{1}{3}\,\mathrm{Re}(\mu(a,b,c,x) + \mu(b,c,a,x) + \mu(c,b,a,x))$$

$(\mathrm{Re}(\alpha) = \frac{1}{2}(\alpha + \bar{\alpha})$ and $\mathrm{Im}(\alpha) = \frac{1}{2}(\bar{\alpha} - \alpha))$.

LEMMA 18. Two $\mathbb{R}$-J*-algebras j_1 and j_2 are $\mathbb{R}$-J*-isomorphic if, and only if, there exist: a $\mathbb{R}$-basis $\mathcal{B}_1$ of j_1, a $\mathbb{R}$-basis $\mathcal{B}_2$ of j_2 and a bijective mapping $L : \mathcal{B}_1 \to \mathcal{B}_2$ such that

(1) $\lambda_1(a,b,c,x) = \lambda_2(L(a),L(b),L(c),L(x))$ whenever $a,b,c,x \in \mathcal{B}_1$,

where λ_1 and λ_2 are the structure coefficients of $\mathcal{B}_1$ and $\mathcal{B}_2$, respectively. The $\mathbb{R}$-J*-isomorphism is unique and its restriction to $\mathcal{B}_1$ is L.

LEMMA 19. Two $\mathbb{C}$-J*-algebras $\mathcal{J}_1$ and $\mathcal{J}_2$ are $\mathbb{C}$-J*-isomorphic if, and only if, there exist: a $\mathbb{C}$-basis $\mathcal{B}_1$ of $\mathcal{J}_1$, a $\mathbb{C}$-basis $\mathcal{B}_2$ of $\mathcal{J}_2$ and a bijective mapping $L : \mathcal{B}_1 \to \mathcal{B}_2$ such that

$$(2) \quad \mu_1(a,b,c,x) = \mu_2(L(a),L(b);L(c),L(x)) \quad \text{whenever} \quad a,b,c,x \in \mathcal{B}_1$$

where μ_1 and μ_2 are the complex structure coefficients of $\mathcal{B}_1$ and $\mathcal{B}_2$, respectively. The $\mathbb{C}$-J*-isomorphism is unique and its restriction to $\mathcal{B}_1$ is L.

Assume $\mathcal{B}$ is a subset of a $\mathbb{R}$-J*-algebra $\mathcal{J}$ such that $\{abc\} \in V_{\mathbb{R}}(\mathcal{B})$, for $a,b,c \in \mathcal{B}$. Then $V_{\mathbb{R}}(\mathcal{B})$ is a $\mathbb{R}$-J*-sub algebra of $\mathcal{J}$. In particular if $[abc] \in V_{\mathbb{R}}(\mathcal{B})$ for $a,b,c \in \mathcal{B}$, since $\{abc\} = \frac{1}{3}([abc] + [bca] + [cab])$, then $V_{\mathbb{R}}(\mathcal{B})$ is an $\mathbb{R}$-J*-algebra of $\mathcal{J}$.

Similarly, if $\mathcal{B}$ is a subset of a $\mathbb{C}$-J*-algebra $\mathcal{J}$ such that $[abc] \in V_{\mathbb{C}}(\mathcal{B})$ for $a,b,c \in \mathcal{B}$, then $V_{\mathbb{C}}(\mathcal{B})$ is a $\mathbb{C}$-J*-sub algebra of $\mathcal{J}$.

PART 2.

In this section we characterize some real and complex J*-algebras.

Let us denote by $\mathbb{R}$, $\mathbb{C}$ and $\mathbb{H}$ respectively the field of real numbers, the field of complex numbers and the non commutative field of Hamiltonian quaternions. Henceforth, we shall identify $\mathbb{C}^n$ with $\mathbb{R}^{2n}$ and $\mathbb{H}^n$ with $\mathbb{R}^{4n}$. Hence $\mathbb{R}^n$ and $\mathbb{H}^n$ will be considered as $\mathbb{R}$-Hilbert spaces and $\mathbb{C}^n$ either a $\mathbb{C}$-Hilbert space or a $\mathbb{R}$-Hilbert space depending upon the necessity. Moreover we shall naturally identify the $m \times n$ matrices with entries in $\mathbb{K}$ ($\mathbb{K}$ is $\mathbb{R}$, $\mathbb{C}$ or $\mathbb{H}$), $M(m,n;\mathbb{K})$, with the $\mathbb{R}$-linear mappings of the $\mathbb{R}$-Hilbert space $\mathbb{K}^n$ into the $\mathbb{R}$-Hilbert space $\mathbb{K}^m$ defined by:

$$(1) \quad \|a_{ij}\| : (x_1,\ldots,x_n) \to (\Sigma_r a_{1r} x_r,\ldots,\Sigma_r a_{mr} x_r) \quad (r = 1,\ldots,n).$$

(note that $M(m,n;\mathbb{C})$ can be naturally identified with the set of all $\mathbb{C}$-linear mappings of the $\mathbb{C}$-Hilbert space $\mathbb{C}^n$ into the $\mathbb{C}$-Hilbert space $\mathbb{C}^m$).

These identifications imply that if $\|a_{ij}\| \in M(m,n;\mathbb{K})$ then $\|a_{ij}\|^* = \|b_{rs}\| \in M(n,m;\mathbb{K})$, where $b_{rs} = \bar{a}_{sr}$, $r = 1,\ldots,n$ and $s = 1,\ldots,m$ ($\bar{a}_{sr}$ is the conjugate in $\mathbb{K}$ of a_{sr}).

Now, let us define some useful $\mathbb{K}$-linear mappings between matrix spaces.

Let $\Lambda,\Theta,\Xi,\Phi,\Omega : M(n;\mathbb{K}) \to M(2n;\mathbb{K})$ ($M(n;\mathbb{K})$ denotes $M(n,n;\mathbb{K})$) be the mappings defined respectively by:

A) $\quad \Lambda(X) = \begin{Vmatrix} 0 & X \\ X & 0 \end{Vmatrix}$

B) $\quad \Theta(X) = \begin{Vmatrix} 0 & X \\ -X & 0 \end{Vmatrix}$

C) $\quad \Xi(X) = \begin{Vmatrix} X & 0 \\ 0 & X \end{Vmatrix}$

D) $\quad \Phi(X) = \begin{Vmatrix} X & 0 \\ 0 & -X \end{Vmatrix}$

E) set $X = \|x_{ij}\|$; the matrix $\Omega(X) = \|b_{rs}\|$ has the entries given by:

$$b_{2i\ 2j} = b_{2i-1\ 2j-1} = x_{ij} \qquad i,j = 1,\dots,n$$

$$b_{2i\ 2j-1} = b_{2i-1\ 2j} = 0 \qquad i,j = 1,\dots,n.$$

LEMMA 20. For $n = 1,\dots,n$, let $\Lambda,\Theta,\Xi,\Phi,\Omega : M(n,\mathbb{K}) \to M(2n,\mathbb{K})$ be the mappings defined above. We have:

A) Λ,Θ,Ξ,Φ and Ω are $\mathbb{K}$-linear;

B) $\Omega\Lambda = \Lambda\Omega$, $\Omega\Theta = \Theta\Omega$, $\Omega\Xi = \Xi\Omega$ and $\Omega\Phi = \Phi\Omega$;

C) $\Xi(X\ Y) = \Xi(X)\ \Xi(Y) = \Phi(X)\ \Phi(Y) = \Lambda(X)\ \Lambda(Y) = -\Theta(X)\ \Theta(Y)$ whenever $X, Y \in M(n;\mathbb{K})$;

D) $\Theta(X) = \Theta(I_n)\ \Lambda(X) = -\Lambda(X)\ \Phi(I_n)$ whenever $x \in M(n;\mathbb{K})$; (I_n is the identity in $M(n;\mathbb{K})$)

E) $\Phi(X) = \Phi(I_n)\ \Xi(X) = \Xi(X)\ \Phi(I_n)$ whenever $X \in M(n;\mathbb{K})$;

F) $\Omega(X^*) = \Omega(X)^*$, $\Lambda(X^*) = \Lambda(X)^*$, $\Theta(X^*) = -\Theta(X)^*$, $\Xi(X^*) = \Xi(X)^*$, $\Phi(X^*) = \Phi(X)^*$ whenever $X \in M(n;\mathbb{K})$.

PROOF. It is trivial computation.

LEMMA 21. Let $n = 1,\dots$; there exist n matrices $A_1,\dots A_n \in M(2^{n-1};\mathbb{R})$ such that:

A) $A_iA_j^* + A_jA_i^* = A_i^*A_j + A_j^*A_i = 0$ whenever $1 \leqslant i < j \leqslant n$;

B) $A_iA_i^* = A_i^*A_i = I_{2^{n-1}}$ whenever $i = 1,\dots,n$.

PROOF. Set $A_1 = I_{2^{n-1}}$ and $A_i = \Omega^{i-2}\Phi^{n-i}\Theta(I_1)$, whenever $i = 2,\dots,n$.

A) and B) are now trivial consequences of lemma 20.

Let us denote by U_n the subset of $M(2^{n-1};\mathbb{R})$

$$U_n = \{A_1,\dots,A_n\}.$$

<u>COROLLARY 4</u>. Let $\mathcal{U}_n = \mathcal{V}_{\mathbb{R}}(U_n) \subset \mathcal{M}(2^{n-1};\mathbb{R})$, for $n = 1,\ldots$.

A) $\mathcal{U}_n$ is a $\mathbb{R}$-J*-algebra;

B) U_n is a basis of $\mathcal{U}_n$ and has the following real structure formulae:

$\{A_1 A_i A_1\} = A_i$ for $i = 1,\ldots,n$;

$\{A_i A_1 A_j\} = \frac{1}{3} A_j$ for $i,j = 1,\ldots,n$ with $i \neq j$;

$\{A_i A_j A_r\} = 0$ for $i,j,r = 1,\ldots,n$ and i,j,r are three distinct integers.

<u>PROOF</u>. Linear independence of the set $\{A_1,\ldots,A_n\}$ is a trivial consequence of the definitions of A_i given in the proof of lemma 21. The structure formulae given in B) are obtained by lemma 20. A) is a trivial consequence of the note given at the end of part 1 in view of structure formulae given in B).

<u>COROLLARY 5</u>. Let $\mathcal{J}$ be a $\mathbb{R}$-J*-algebra. If $\mathcal{J}$ has a non zero, idempotent, irreducible element a such that δ_a is the identity on $\mathcal{J}$ and $\dim\mathcal{J} = n > 0$, then $\mathcal{J}$ is $\mathbb{R}$-J*-isomorphic to $\mathcal{U}_n$. In particular:

A) $\mathcal{M}(1;\mathbb{R})$ is $\mathbb{R}$-J*-isomorphic to $\mathcal{U}_1$;

B) $\mathcal{M}(1;\mathbb{C})$ is $\mathbb{R}$-J*-isomorphic to $\mathcal{U}_2$;

C) $\mathcal{M}(1;\mathbb{H})$ is $\mathbb{R}$-J*-isomorphic to $\mathcal{U}_4$.

<u>PROOF</u>. The main part of this corollary is a consequence of lemma 11 and lemma 18.

A) is trivial. B) is a consequence of lemma 18 where $\sigma : \{\|1\|, \|i\|\} \to \{A_1,A_2\}$ is an arbitrary bijective mapping.

C) is a consequence of lemma 18 too where $\sigma : \{\|1\|, \|i\|, \|j\|, \|k\|\} \to \{A_1,A_2,A_3,A_4\}$ is an arbitrary bijective mapping.

Let H and K denote respectively the matrices $\begin{Vmatrix} 1 & 0 \\ 0 & 0 \end{Vmatrix}$ and $\begin{Vmatrix} 0 & 0 \\ 0 & 1 \end{Vmatrix}$, and consider the elements of $M(2^s;\mathbb{R})$ defined by:

$T^s_s = \Omega^{s-1}(H)$, whenever $s \geq 1$;

$\tilde{T}^s_s = -\Omega^{s-1}(K)$, whenever $s \geq 1$;

$T^s_p = \Omega^{p-s}\Lambda^{s-p-1}\left(\begin{Vmatrix} 0 & H \\ -K & 0 \end{Vmatrix}\right)$, whenever $s \geq 2$ and $p = 1,\dots,s-1$;

$\tilde{T}^s_p = T^s_p{}^*$, whenever $s \geq 2$ and $p = 1,\dots,s-1$;

$\approx{T}^s_p = T^s_p$, whenever $1 \leq p \leq s$.

Moreover, let us denote

$T_{2n} = \{T^n_p, \tilde{T}^n_p : 1 \leq p \leq n\} \subset M(2^n;\mathbb{R}) \subset M(2^n;\mathbb{C})$;

$I_{2n} = V_{\mathbb{R}}(T_{2n})$;

$I^{\mathbb{C}}_{2n} = V_{\mathbb{C}}(T_{2n})$.

LEMMA 22. $I^{\mathbb{C}}$ is a $\mathbb{C}$-J*-sub algebra of $M(2^n;\mathbb{C})$ of $\mathbb{C}$-dimension 2n. T_{2n} is a $\mathbb{C}$-basis of $I^{\mathbb{C}}_{2n}$ with complex structure formulae:

A) $[xxx] = x$, whenever $x \in T_{2n}$;

B) $[xxy] = \frac{1}{2}y$, whenever $x, y \in T_{2n}$ such that $x, y, \tilde{x}, \tilde{y}$ are distinct elements;

C) $[xyx] = 0$, whenever $x, y \in T_{2n}$ such that $x, y, \tilde{x}, \tilde{r}$ are distinct elements;

D) $[\tilde{x}xy] = 0$, whenever $x, y \in T_{2n}$ such that $x, y, \tilde{x}, \tilde{y}$ are distinct elements;

E) $[xy\tilde{x}] = \frac{1}{2}y$, whenever $x, y \in T_{2n}$ such that $x, y, \tilde{x}, \tilde{y}$ are distinct elements;

F) $[xx\tilde{x}] = 0$, whenever $x \in T_{2n}$;

G) $[x\tilde{x}x] = 0$, whenever $x \in T_{2n}$;

H) $[xyz] = 0$, whenever $x, y, z \in T_{2n}$ such that $x, y, z, \tilde{x}, \tilde{y}, \tilde{z}$ are distinct elements.

PROOF. A), ..., H) are obtained trivially from the following relations:

i) $x^*\tilde{x} = \tilde{x}^*x = 0\ x\tilde{x}^* = \tilde{x}x^* = 0$, whenever $x \in T_{2n}$;

ii) $x^*y + \tilde{y}^*\tilde{x} = 0$, $xy^* + \widetilde{yx}^* = 0$, whenever $x, y \in T_{2n}$ such that $x, y, \tilde{x}, \tilde{y}$ are distinct elements;

iii) $T^n_n{}^* T^n_n = T^n_n T^n_n{}^* = T^n_n$;

iv) $\tilde{T}^n_n{}^* \tilde{T}^n_n = \tilde{T}^n_n \tilde{T}^n_n{}^* = T^n_n$;

v) $T^n_p{}^* T^n_p = \tilde{T}^n_p \tilde{T}^n_p{}^* = \Omega^{p-1}\Xi^{n-p-1}\left(\begin{Vmatrix} K & 0 \\ 0 & H \end{Vmatrix}\right)$, whenever $1 \leqslant p < n$;

vi) $T^n_p T^n_p{}^* = \tilde{T}^n_p{}^* \tilde{T}^n_p = \Omega^{p-1}\Xi^{n-p-1}\left(\begin{Vmatrix} H & 0 \\ 0 & K \end{Vmatrix}\right)$, whenever $1 \leqslant p < n$;

Since we have: $T^n_n{}^* \tilde{T}^n_n = -\Omega^{n-1}(H)\Omega^{n-1}(K) = -\Omega^{n-1}(HK) = 0$;

$\tilde{T}^n_n{}^* T^n_n = T^n_n \tilde{T}^n_n{}^* = \tilde{T}^n_n T^n_n{}^* = 0$ (proved analogously);

$T^n_p{}^* \tilde{T}^n_p = \Omega^{p-1}\Lambda^{n-p-1}\left(\begin{Vmatrix} 0 & -K \\ H & 0 \end{Vmatrix}\right)$.

$\Omega^{p-1}\Lambda^{n-p-1}\left(\begin{Vmatrix} 0 & -K \\ H & 0 \end{Vmatrix}\right) = \Omega^{p-1}\Xi^{n-p-1}\left(\begin{Vmatrix} 0 & -K \\ H & 0 \end{Vmatrix}\begin{Vmatrix} 0 & -K \\ H & 0 \end{Vmatrix}\right) = 0$, whenever $1 \leqslant p < n$; $\tilde{T}^n_p T^n_p{}^* = T^n_p{}^* T^n_p = T^n_p \tilde{T}^n_p{}^* = 0$ (proved analogously).

i) is trivially verified.

Whenever $1 \leqslant p < n$, we have:

$$T^n_n{}^* T^n_p + \tilde{T}^n_p{}^* \tilde{T}^n_n = \Omega^{n-1}(H)\,\Omega^{p-1}\Lambda^{n-p-1}\left(\begin{Vmatrix} 0 & H \\ -K & 0 \end{Vmatrix}\right) -$$

$$- \Omega^{p-1}\Lambda^{n-p-1}\left(\begin{Vmatrix} 0 & H \\ -K & 0 \end{Vmatrix}\right)\Omega^{n-1}(K) = \Omega^{p-1}\left(\Omega^{n-p}(H)\,\Lambda^{n-p-1}\left(\begin{Vmatrix} 0 & H \\ -K & 0 \end{Vmatrix}\right) -\right.$$

$$\left. - \Lambda^{n-p-1}\left(\begin{Vmatrix} 0 & H \\ -K & 0 \end{Vmatrix}\right)\Omega^{n-p}(K)\right) = \Omega^{p-1}\left(\begin{Vmatrix} I & 0 \\ 0 & 0 \end{Vmatrix}\begin{Vmatrix} 0 & P \\ R & 0 \end{Vmatrix} -\right.$$

$$\left. - \begin{Vmatrix} 0 & P \\ R & 0 \end{Vmatrix}\begin{Vmatrix} 0 & 0 \\ 0 & I \end{Vmatrix}\right) = 0$$

(we have denoted $P = R = \Lambda^{n-p-2}\left(\begin{Vmatrix} 0 & H \\ -K & 0 \end{Vmatrix}\right)$ whenever $n-p-2 \geqslant 0$, $P = H$ and $R = -K$ if $n-p-2 < 0$).

With the same argument we can prove that, whenever $1 \leq p < n$, we have $T^n_p * T^n_n + \tilde{T}^n_n * \tilde{T}^n_p = T^n_p T^n_n{}^* + \tilde{T}^n_n \tilde{T}^n_p{}^* = T^n_n T^n_p{}^* + \tilde{T}^n_p \tilde{T}^n_n{}^* = 0$.

If $1 \leq p < q \leq n$, we have:

$$T^n_p * T^n_q + \tilde{T}^n_q * \tilde{T}^n_p = \Omega^{p-1}\Lambda^{n-p-1}\left(\begin{Vmatrix} 0 & -K \\ H & 0 \end{Vmatrix}\right)\Omega^{q-1}\Lambda^{n-q-1}\left(\begin{Vmatrix} 0 & H \\ -K & 0 \end{Vmatrix}\right) +$$

$$+ \Omega^{q-1}\Lambda^{n-q-1}\left(\begin{Vmatrix} 0 & H \\ -K & 0 \end{Vmatrix}\right)\Omega^{p-1}\Lambda^{n-p-1}\left(\begin{Vmatrix} 0 & -K \\ H & 0 \end{Vmatrix}\right) =$$

$$= \Omega^{p-1}\Lambda^{n-q-1}\left(\Lambda^{q-p}\left(\begin{Vmatrix} 0 & -K \\ H & 0 \end{Vmatrix}\right)\Omega^{q-p}\left(\begin{Vmatrix} 0 & H \\ -K & 0 \end{Vmatrix}\right) + \Omega^{q-p}\left(\begin{Vmatrix} 0 & H \\ -K & 0 \end{Vmatrix}\right).\right.$$

$$\left. . \Lambda^{q-p}\left(\begin{Vmatrix} 0 & -K \\ H & 0 \end{Vmatrix}\right)\right) = \Omega^{p-1}\Lambda^{n-q-1}\left(\begin{Vmatrix} R & 0 \\ 0 & S \end{Vmatrix}\right) = 0$$

(where we have denoted $R = -\Lambda^{q-p-1}\left(\begin{Vmatrix} 0 & -K \\ H & 0 \end{Vmatrix}\right)\Omega^{q-p}(K) +$ $+\Omega^{q-p}(H)\Lambda^{q-p-1}\left(\begin{Vmatrix} 0 & -K \\ H & 0 \end{Vmatrix}\right)$ and $S = \Lambda^{q-p-1}\left(\begin{Vmatrix} 0 & -K \\ H & 0 \end{Vmatrix}\right)\Omega^{q-p}(H) -$ $- \Omega^{q-p}(K)\Lambda^{q-p-1}\left(\begin{Vmatrix} 0 & -K \\ H & 0 \end{Vmatrix}\right)$ with the convention $\Lambda^0\left(\begin{Vmatrix} 0 & -K \\ H & 0 \end{Vmatrix}\right) = \begin{Vmatrix} 0 & -K \\ H & 0 \end{Vmatrix}$). The other relations necessary to complete the proof of ii) are obtained in the same way. iii) and iv) can be proved with a direct computation too. The proof is now a consequence of the considerations made at the end of part 1 and the linear independence of T_{2n}.

<u>LEMMA 23</u>. I_{2n} is a $\mathbb{R}$-J*-sub algebra of $M(2^n;\mathbb{R})$ of $\mathbb{R}$-dimension $2n$. T_{2n} is an $\mathbb{R}$-basis of I_{2n} with real structure formulae

A) $\{xxx\} = x$, whenever $x \in T_{2n}$;

B) $\{xxy\} = \frac{1}{3} y$, whenever $x, y \in T_{2n}$ such that $x,y,\tilde{x},\tilde{y}$ are distinct elements;

C) $\{x\tilde{x}y\} = -\frac{1}{6}\tilde{y}$, whenever $x, y \in T_{2n}$ such that $x,y,\tilde{x},\tilde{y}$ are distinct elements;

D) $\{xx\tilde{x}\} = 0$, whenever $x \in T_{2n}$;

E) $\{xyz\} = 0$, whenever $x,y,z \in T_{2n}$ such that $x,y,z,\tilde{x},\tilde{y},\tilde{z}$ are distinct elements.

<u>PROOF</u>. It is a consequence of lemma 22 and the note at the end of part 1.

Let us denote $U_0 = \{\|0\|\}$ and, whenever $p = 1,2,\dots,$ U_p the set defined at the end of lemma 21. Moreover, whenever $n \geq 1$ and $p \geq 0$, set

$$T_{2n\,p} = \Omega^p(T_{2n}) \cup \Lambda^n(\{\begin{Vmatrix} 0 & R \\ R^* & 0 \end{Vmatrix} : R \in U_p\}) - \{0\},$$

and $\Omega^p(x)^\sim = \Omega^p(\tilde{x})$, whenever $x \in T_{2n}$.

<u>LEMMA 24</u>. $I_{2n\,p} = V_{\mathbb{R}}(T_{2n\,p}) \subset M(2^{n+p};\mathbb{R})$ is a $\mathbb{R}$-J*-sub algebra of $M(2^{n+p};\mathbb{R})$ of dimension $2n + p$. $T_{2n\,p}$ is a basis of $I_{2n\,p}$ and has the following real structure formulae:

A) $\{xxx\} = x$, whenever $x \in \Omega^p(T_{2n})$;

B) $\{xxy\} = \frac{1}{3}\, y$, whenever $x, y \in \Omega^p(T_{2n})$ such that $x,y,\tilde{x},\tilde{y}$ are distinct elements;

C) $\{x\tilde{x}y\} = -\frac{1}{6}\,\tilde{y}$, whenever $x, y \in \Omega^p(T_{2n})$ such that $x,y,\tilde{x},\tilde{y}$, are distinct elements;

D) $\{\tilde{x}xx\} = 0$, whenever $x \in \Omega^p(T_{2n})$;

E) $\{xyz\} = 0$, whenever $x,y,z \in \Omega^p(T_{2n})$ such that $x,y,z,\tilde{x},\tilde{y},\tilde{z}$, are distinct elements;

F) $\{xxx\} = x$, whenever $x \in \Lambda^n(\{\begin{Vmatrix} 0 & R \\ R^* & 0 \end{Vmatrix} : R \in U_p\}) - \{0\} = W_{2n\,p}$;

G) $\{xxy\} = \frac{1}{3}\, y$, whenever $x, y \in W_{2n\,p}$ with $x \neq y$;

H) $\{xyz\} = 0$, whenever $x,y,z \in W_{2n\,p}$ such that x,y,z are distinct elements;

I) $\{xxy\} = \frac{1}{3}\, y$, whenever $x \in \Omega^p(T_{2n})$ and $y \in W_{2n\,p}$;

J) $\{xxy\} = \frac{2}{3}\, y - \frac{1}{3}\,\tilde{y}$, whenever $x \in W_{2n\,p}$ and $y \in \Omega^p(T_{2n})$;

K) $\{x\tilde{x}y\} = -\frac{1}{6}\, y$, whenever $x \in \Omega^p(T_{2n})$ and $y \in W_{2n\,p}$;

L) $\{xyz\} = 0$, whenever $x,y \in \Omega^p(T_{2n})$ such that $x,y,\tilde{x},\tilde{y}$ are distinct elements and $z \in W_{2n\,p}$;

M) $\{xyz\} = 0$, whenever $x, y \in W_{2n\,p}$ with $x \neq y$ and $x \in \Omega^p(T_{2n})$.

PROOF. Since $\Omega(\{xyz\}) = \{\Omega(x)\Omega(y)\Omega(z)\}$ and $\Lambda(\{xyz\}) = \{\Lambda(x)\Lambda(y)\Lambda(z)\}$, for $x,y,z \in M(s;\mathbb{K})$ (see lemma 20), A), ..., E) follow from lemma 23 and F), G), H) by corollary 4. So we have only to prove I), ..., M). The proof breaks into several particular cases. Here we give only a hint and leave detailed calculations to the reader.

Using the notation established in the previous lemmas, we have:

$$3\{\Omega^p(H)\Omega^p(H)\Lambda^n(\begin{Vmatrix}0 & R\\ R^* & 0\end{Vmatrix})\} = \Omega^{n+p-1}(H)\Lambda^n(\begin{Vmatrix}0 & R\\ R^* & 0\end{Vmatrix}) +$$

$$+ \Lambda^n(\begin{Vmatrix}0 & R\\ R^* & 0\end{Vmatrix})\Omega^{n+p-1}(H) + \Omega^{n+p-1}(H)\Lambda^n(\begin{Vmatrix}0 & R\\ R^* & 0\end{Vmatrix})\Omega^{n+p-1}(H) =$$

$$= \begin{Vmatrix}I & 0\\ 0 & 0\end{Vmatrix}\begin{Vmatrix}0 & M\\ N & 0\end{Vmatrix} + \begin{Vmatrix}0 & M\\ N & 0\end{Vmatrix}\begin{Vmatrix}I & 0\\ 0 & 0\end{Vmatrix} + \begin{Vmatrix}I & 0\\ 0 & 0\end{Vmatrix}\begin{Vmatrix}0 & M\\ N & 0\end{Vmatrix}\begin{Vmatrix}I & 0\\ 0 & 0\end{Vmatrix} =$$

$$= \Lambda^n(\begin{Vmatrix}0 & R\\ R^* & 0\end{Vmatrix}).$$

(We have set $M = N = \Lambda^{n-1}(\begin{Vmatrix}0 & R\\ R^* & 0\end{Vmatrix})$, whenever $n \geqslant 2$ and $M = R$, $N = R^*$, whenever $n \leqslant 1$). In the same way we can prove

$$3\{\Lambda^p(K)\Omega^p(K)\Lambda^n(\begin{Vmatrix}0 & R\\ R^* & 0\end{Vmatrix})\} = \Lambda^n(\begin{Vmatrix}0 & R\\ R^* & 0\end{Vmatrix}).$$

Let now $1 \leqslant h < n$. We have:

$$3\{\Omega^p(T^n_h)\Omega^p(T^n_h)\Lambda^n(\begin{Vmatrix}0 & R\\ R^* & 0\end{Vmatrix})\} =$$

$$= \Lambda^{n-h-1}\Omega^{h+p-1}(\begin{Vmatrix}0 & H\\ -K & 0\end{Vmatrix})\Lambda^{n-h-1}\Omega^{h+p-1}(\begin{Vmatrix}0 & -K\\ H & 0\end{Vmatrix})\Lambda^n(\begin{Vmatrix}0 & R\\ R^* & 0\end{Vmatrix}) +$$

$$+ \Lambda^n(\begin{Vmatrix}0 & R\\ R^* & 0\end{Vmatrix})\Lambda^{n-h-1}\Omega^{h+p-1}(\begin{Vmatrix}0 & -K\\ H & 0\end{Vmatrix})\Lambda^{n-h-1}\Omega^{h+p-1}(\begin{Vmatrix}0 & H\\ -K & 0\end{Vmatrix}) +$$

$$+ \Lambda^{n-h-1}\Omega^{h+p-1}(\begin{Vmatrix}0 & H\\ -K & 0\end{Vmatrix})\Lambda^n(\begin{Vmatrix}0 & R\\ R^* & 0\end{Vmatrix})\Lambda^{n-h-1}\Omega^{h+p-1}(\begin{Vmatrix}0 & H\\ -K & 0\end{Vmatrix}) +$$

$$+ \Lambda^{n-h-1}(\begin{Vmatrix}I & 0 & 0 & 0\\ 0 & 0 & 0 & 0\\ 0 & 0 & 0 & 0\\ 0 & 0 & 0 & I\end{Vmatrix}\begin{Vmatrix}0 & \Lambda^h(\begin{Vmatrix}0 & R\\ R^* & 0\end{Vmatrix})\\ \Lambda^h(\begin{Vmatrix}0 & R\\ R^* & 0\end{Vmatrix}) & 0\end{Vmatrix} +$$

$$+ \begin{Vmatrix} 0 & \Lambda^h\left(\begin{Vmatrix} 0 & R \\ R^* & 0 \end{Vmatrix}\right) \\ \Lambda^h\left(\begin{Vmatrix} 0 & R \\ R^* & 0 \end{Vmatrix}\right) & 0 \end{Vmatrix} \begin{Vmatrix} 0 & 0 & 0 & 0 \\ 0 & I & 0 & 0 \\ 0 & 0 & I & 0 \\ 0 & 0 & 0 & 0 \end{Vmatrix} +$$

$$+ \begin{Vmatrix} 0 & 0 & I & 0 \\ 0 & 0 & 0 & 0 \\ 0 & 0 & 0 & 0 \\ 0 & -I & 0 & 0 \end{Vmatrix} \begin{Vmatrix} 0 & \Lambda^h\left(\begin{Vmatrix} 0 & R \\ R^* & 0 \end{Vmatrix}\right) \\ \Lambda^h\left(\begin{Vmatrix} 0 & R \\ R^* & 0 \end{Vmatrix}\right) & 0 \end{Vmatrix} \begin{Vmatrix} 0 & 0 & I & 0 \\ 0 & 0 & 0 & 0 \\ 0 & 0 & 0 & 0 \\ 0 & -I & 0 & 0 \end{Vmatrix}) =$$

$$= \Lambda^n\left(\begin{Vmatrix} 0 & R \\ R^* & 0 \end{Vmatrix}\right) + \Lambda^{n-h-1}\left(\begin{Vmatrix} 0 & \begin{Vmatrix} I & 0 \\ 0 & 0 \end{Vmatrix} \Lambda^h\left(\begin{Vmatrix} 0 & R \\ R & 0 \end{Vmatrix}\right) \begin{Vmatrix} I & 0 \\ 0 & 0 \end{Vmatrix} \\ \begin{Vmatrix} 0 & 0 \\ 0 & I \end{Vmatrix} \Lambda^h\left(\begin{Vmatrix} 0 & R \\ R^* & 0 \end{Vmatrix}\right) \begin{Vmatrix} 0 & 0 \\ 0 & I \end{Vmatrix} & 0 \end{Vmatrix}\right) =$$

$$= \Lambda^n\left(\begin{Vmatrix} 0 & R \\ R^* & 0 \end{Vmatrix}\right).$$

In the same way we can prove

$$3\, \Omega^p(T^n{}_h)\Omega^p(T^n{}_h)\Lambda^n\left(\begin{Vmatrix} 0 & R \\ R^* & 0 \end{Vmatrix}\right) = \Lambda^n\left(\begin{Vmatrix} 0 & R \\ R^* & 0 \end{Vmatrix}\right).$$

The proof of I) is now complete. The proof of J), ..., M) is left to the reader.

<u>COROLLARY 6</u>. The $\mathbb{R}$-J*-algebra of symmetric real matrices of order 2 is $\mathbb{R}$-J*-isomorphic to $I_{2\,1}$. The $\mathbb{R}$-J*-algebra of complex Hermitian matrices of order 2 is $\mathbb{R}$-J*-isomorphic to $I_{2\,2}$. The $\mathbb{R}$-J*-algebra of quaternionic Hermitian matrices of order 2 is $\mathbb{R}$-J*-isomorphic to $I_{2\,4}$.

<u>PROOF</u>. A trivial consequence of lemma 18.

Observe that I_{2n} is $\mathbb{R}$-J*-isomorphic to $I_{2n\,0}$.

<u>LEMMA 25</u>. Let $n = 1,2,\ldots$ and $p = 0,1$. $I^{\mathbb{C}}_{2n\,p} = V_{\mathbb{C}}(T_{2n\,p}) \subset M(2^{n+p};\mathbb{C})$ is a $\mathbb{C}$-J*-sub algebra of $M(2^{n+p};\mathbb{C})$ of $\mathbb{C}$-dimension $2n+p$. $T_{2n\,p}$ is a $\mathbb{C}$-basis of $I^{\mathbb{C}}_{2n\,p}$ and has the following complex structure formulae:

A) $[xxx] = x$, whenever $x \in \Omega^p(T_{2n})$;

B) $[xxy] = \frac{1}{2}\, y$, whenever $x, y \in \Omega^p(T_{2n})$ such that $x, y, \tilde{x}, \tilde{y}$ are distinct elements;

C) $[xyx] = 0$, whenever $x, y \in \Omega^p(T_{2n})$ such that $x, y, \tilde{x}, \tilde{y}$ are distinct elements;

D) $[x\tilde{x}y] = 0$, whenever $x, y \in \Omega^p(T_{2n})$ such that $x, y, \tilde{x}, \tilde{y}$ are distinct elements;

E) $[xy\tilde{x}] = -\frac{1}{2}\tilde{y}$, whenever $x, y \in \Omega^p(T_{2n})$ such that $x, y, \tilde{x}, \tilde{y}$ are distinct elements;

F) $[xx\tilde{x}] = 0$, whenever $x \in \Omega^p(T_{2n})$;

G) $[x\tilde{x}x] = 0$, whenever $x \in \Omega^p(T_{2n})$;

H) $[xyz] = 0$, whenever $x, y, z \in \Omega^p(T_{2n})$ such that $x, y, z, \tilde{x}, \tilde{y}, \tilde{z}$ are distinct elements;

I) $[xxx] = x$, whenever $x \in \Lambda^n(\{\begin{Vmatrix} 0 & R \\ R^* & 0 \end{Vmatrix} : R \in U_p\}) - \{0\} = W_{2n\,p}$;

J) $[xxy] = \frac{1}{2}y$, whenever $x \in \Omega^p(T_{2n})$ and $y \in W_{2n\,p}$;

K) $[xyx] = 0$, whenever $x \in \Omega^p(T_{2n})$ and $y \in W_{2n\,p}$;

L) $[xxy] = y$, whenever $x \in W_{2n\,p}$ and $y \in \Omega^p(T_{2n})$;

M) $[xyx] = -\frac{1}{2}\tilde{y}$, whenever $x \in W_{2n\,p}$ and $y \in \Omega^p(T_{2n})$;

N) $[x\tilde{x}y] = 0$, whenever $x \in \Omega^p(T_{2n})$ and $y \in W_{2n\,p}$;

O) $[xy\tilde{x}] = -\frac{1}{2}y$, whenever $x \in \Omega^p(T_{2n})$ and $y \in W_{2n\,p}$;

P) $[xyz] = 0$, whenever $x, y \in \Omega^p(T_{2n})$ such that $x, y, \tilde{x}, \tilde{y}$ are distinct elements and $z \in W_{2n\,p}$;

Q) $[xyz] = 0$, whenever $x, z \in \Omega^p(T_{2n})$ such that $x, z, \tilde{x}, \tilde{z}$ are distinct elements and $y \in W_{2n\,p}$.

PROOF. A), ..., H) are consequences of lemma 22. I) is a consequence of the definition of U_p. J), ..., Q) are proved considering a number of particular cases, and the details are left to the reader.

Observe that if $p \geq 2$:

$$[\Omega^p(T^n{}_n)\Lambda^n(\begin{Vmatrix} 0 & A_1 \\ A_1^* & 0 \end{Vmatrix})\Lambda^n(\begin{Vmatrix} 0 & A_2 \\ A_2^* & 0 \end{Vmatrix})] \notin V_{\mathbb{C}}(T_{2n\,p})$$

(here we have used the notation established in the previous lemmas) and hence $V_{\mathbb{C}}(T_{2n\,p})$ is not a $\mathbb{C}$-J*-sub algebra of $M(2^{n+p};\mathbb{C})$ if $p \geq 2$.

Let $n = 1,2,\ldots$ and $1 \leq p \leq n$. Define:

$$E^n_p = \Omega^{n-p}\Lambda^{p-1}\left(\begin{Vmatrix} 0 & 1 \\ -1 & 0 \end{Vmatrix}\right) \in M(2^n;\mathbb{C});$$

$$E_n = \{E^n_1, E^n_2, \ldots, E^n_n\},$$

$$\mathcal{E}_n = V_{\mathbb{C}}(E_n).$$

<u>LEMMA 26</u>. $\mathcal{E}_n = V_{\mathbb{C}}(E_n)$ is a $\mathbb{C}$-J*-sub algebra of $M(2^n;\mathbb{C})$ of $\mathbb{C}$-dimension n.

E_n is a $\mathbb{C}$-basis of $\mathcal{E}_n$ with complex structure formulae:

A) $[xxx] = x$, whenever $x \in E_n$;

B) $[xxy] = y$, whenever $x, y \in E_n$ with $x \neq y$;

C) $[xyx] = -y$, whenever $x, y \in E_n$ with $x \neq y$;

D) $[xyz] = 0$, whenever x,y,z are distinct elements of E_n.

$\mathcal{E}_n$ is $\mathbb{C}$-J*-isomorphic to $I_{2m\,n-2m}$ where $m = [\frac{n}{2}]$ (the highes integer $\leq [\frac{n}{2}]$).

<u>PROOF</u>. First of all we prove the following relations:

i) $E^n_p E^n_p = -I_{2^n}$ whenever $1 \leq p \leq n$;

ii) $E^{n*}_p = -E^n_p$, whenever $1 \leq p \leq n$;

iii) $E^n_p E^n_p + E^n_q E^n_p = 0$, whenever $1 \leq p < q \leq n$.

i) and ii) are immediate consequences of the definition of E^n_p and lemma 20. Using again lemma 20, we have:

$$E^n_p E^n_p + E^n_q E^n_p = \Omega^{n-p}\Lambda^{p-1}\left(\begin{Vmatrix} 0 & 1 \\ -1 & 0 \end{Vmatrix}\right)\Omega^{n-q}\Lambda^{q-1}\left(\begin{Vmatrix} 0 & 1 \\ -1 & 0 \end{Vmatrix}\right) +$$

$$+ \Omega^{n-q}\Lambda^{q-1}(\begin{Vmatrix} 0 & 1 \\ -1 & 0 \end{Vmatrix})\Omega^{n-p}\Lambda^{p-1}(\begin{Vmatrix} 0 & 1 \\ -1 & 0 \end{Vmatrix}) =$$

$$= \Omega^{n-q}\Xi^{p-1}(\Omega^{q-p}(\begin{Vmatrix} 0 & 1 \\ -1 & 0 \end{Vmatrix})\Lambda^{q-p}(\begin{Vmatrix} 0 & 1 \\ -1 & 0 \end{Vmatrix}) + \Lambda^{q-p}(\begin{Vmatrix} 0 & 1 \\ -1 & 0 \end{Vmatrix})\Omega^{q-p}(\begin{Vmatrix} 0 & 1 \\ -1 & 0 \end{Vmatrix}))$$

$$= \Omega^{n-q}\Lambda^{p-1}(\begin{Vmatrix} 0 & I \\ -I & 0 \end{Vmatrix} \begin{Vmatrix} 0 & \Lambda^{q-p-1}(\begin{Vmatrix} 0 & 1 \\ -1 & 0 \end{Vmatrix}) \\ \Lambda^{q-p-1}(\begin{Vmatrix} 0 & 1 \\ -1 & 0 \end{Vmatrix}) & 0 \end{Vmatrix}$$

$$\begin{Vmatrix} 0 & {}^{q-p-1}(\begin{Vmatrix} 0 & 1 \\ -1 & 0 \end{Vmatrix}) \\ {}^{q-p-1}(\begin{Vmatrix} 0 & 1 \\ -1 & 0 \end{Vmatrix}) & 0 \end{Vmatrix} (\begin{Vmatrix} 0 & I \\ -I & 0 \end{Vmatrix}) = 0$$

which proves iii). A), ..., D) are now a consequence of i), ii) and iii).

The second part of the lemma follows since

$$\Omega^p(\{x + \tilde{x}, i(x + \tilde{x}) : x \in T_{2n}\}) \cup \Lambda^n(\{\begin{Vmatrix} 0 & R \\ R & 0 \end{Vmatrix} : R \in U_p\}) - \{0\}$$

is a $\mathbb{C}$-basis of $I^{\mathbb{C}}_{2n\,p}$ and the mapping σ defined by

$\sigma\Omega^p(T^n_r + \tilde{T}^n_r) = E^{2n+p}_{2r}$, whenever $1 \leq r \leq n$

$\sigma\Omega^p(iT^n_r + i\tilde{T}^n_r) = E^{2n+p}_{2r-1}$, whenever $1 \leq r \leq n$;

$\sigma\Lambda^n(\begin{Vmatrix} 0 & 1 \\ 1 & 0 \end{Vmatrix}) = E^{2n+p}_{2n+p}$, whenever $p \neq 0$;

satisfies the hypothesis of lemma 19.

<u>LEMMA 27</u>. The following matrix spaces are $\mathbb{R}$-J*-algebras:

A) $M(m,n;\mathbb{R})$, for $m,n = 1,2,\ldots$;

B) $M(m,n;\mathbb{H})$, for $m,n = 1,2,\ldots$;

C) $S(n;\mathbb{R}) = \{\|b_{ij}\| : \|b_{ij}\| \in M(n;\mathbb{R})$ such that $b_{ij} = b_{ji}$ whenever $1 \leq i, j \leq n\}$;

D) $A(n;\mathbb{R}) = \{\|b_{ij}\| : \|b_{ij}\| \in M(n;\mathbb{R})$ such that $b_{ij} = -b_{ji}$ whenever

$1 \leq i, j \leq n\}$;

E) $H(n;\mathbb{C}) = \{ \|b_{ij}\| : \|b_{ij}\| \in M(n;\mathbb{C})$ such that $b_{ij} = \bar{b}_{ji}$ whenever $1 \leq i, j \leq n\}$;

F) $H(n;\mathbb{H}) = \{ \|b_{ij}\| : \|b_{ij}\| \in M(n;\mathbb{H})$ such that $b_{ij} = \bar{b}_{ji}$ whenever $1 \leq i, j \leq n\}$;

G) $H^*(n;\mathbb{H}) = \{ \|b_{ij}\| : \|b_{ij}\| \in M(n;\mathbb{H})$ such that $b_{ij} = -\bar{b}_{ji}$ whenever $1 \leq i, j \leq n\}$.

The following matrix spaces are $\mathbb{C}$-J*-algebras (and hence $\mathbb{R}$-J*-algebras):

H) $M(m,n;\mathbb{C})$, for $m,n = 1,2,\ldots$;

I) $S(n;\mathbb{C}) = \{ \|b_{ij}\| : \|b_{ij}\| \in M(n;\mathbb{C})$ such that $b_{ij} = b_{ji}$ whenever $1 \leq i, j \leq n\}$;

J) $A(n;\mathbb{C}) = \{ \|b_{ij}\| : \|b_{ij}\| \in M(n;\mathbb{C})$ such that $b_{ij} = -b_{ji}$ whenever $1 \leq i, j \leq n\}$.

PROOF. A, B and H) are trivial. Now let $\|b_{ij}\| \in M(m,n;\mathbb{K})$ (where $\mathbb{K}$ is one of the fields $\mathbb{R}$, $\mathbb{C}$, $\mathbb{H}$). Denote $\|b_{ij}\|^t$ the transpose of $\|b_{ij}\|$ i.e. the matrix $\|c_{rs}\| \in M(n,m,\mathbb{K})$ such that $c_{rs} = b_{sr}$ whenever $1 \leq r, s \leq n$. Moreover denote $\overline{\|b_{ij}\|}$ the conjugated matrix of $\|b_{ij}\|$, i.e. the matrix $\|c_{rs}\| \in M(m,n;\mathbb{K})$ such that $c_{rs} = \bar{b}_{rs}$ whenever $1 \leq r, s \leq n$. We have $\|b_{ij}\|^* = \overline{\|b_{ij}\|}^t$.
Clearly the sets C), ..., G) are $\mathbb{R}$-vector spaces of finite dimension and I) and J) $\mathbb{C}$-vector spaces of finite dimension. So we have only to prove that if x is an element of one of these sets, then xx^*x belongs to the same set. Assume $x \in S(n;\mathbb{K})$ ($\mathbb{K} = \mathbb{C}, \mathbb{R}$). Then $(xx^*x)^t = x^t x^{t*} x^t = xx^*x$ and hence C) and I) are J*-algebras.
Assume $x \in A(n;\mathbb{K})$ ($\mathbb{K} = \mathbb{R}, \mathbb{C}$), it is $(xx^*x)^t = x^t x^{t*} x^t = -xx^*x$ and hence D) and J) are J*-algebras. The proof can now be completed with similar arguments.

PART 3.

In the above section we enumerated some $\mathbb{R}$-J*-algebras and $\mathbb{C}$-J*-algebras (some of them are J*-isomorphic).

Here we shall prove that <u>any finite dimensional indecomposable $\mathbb{R}$-J*-algebra is $\mathbb{R}$-J*-isomorphic to one of the J*-algebras given in part 2</u>. Hence we shall obtain a complete classification of finite dimensional $\mathbb{R}$-J*-algebras.

LEMMA 28. Let a and b be two non zero, idempotent, irreducible, strongly independent elements of a J*-algebra $\mathcal{j}$. Assume $\psi_a\psi_b(\mathcal{j})$ has a non zero, idempotent element c such that c is irreducible in $\psi_a\psi_b(\mathcal{j})$ and $6\{abc\} = \epsilon c$ where $\epsilon \in \{1, -1\}$. Then the $\mathbb{R}$-J*-algebra of $\mathcal{j}$

$$\mathcal{j}' = (\delta_a + \delta_b + \psi_a\psi_b)(\mathcal{j})$$

is $\mathbb{R}$-J*-isomorphic to one of the following $\mathbb{R}$-J*-algebras:

$$I_{2\,n},\ S(2;\mathbb{C}),\ H^*(2;\mathbb{H}).$$

PROOF. The proof that $\mathcal{j}'$ is a J*-algebra is trivial. Let us assume that $\mathcal{j}$ is a J*-algebra of bounded $\mathbb{R}$-linear operators of the $\mathbb{R}$-Hilber space V into the $\mathbb{R}$-Hilbert space W. Let $\mu : \psi_a\psi_b(\mathcal{j}) \to V(V,W)$ be the $\mathbb{R}$-linear mapping ($V(V,W)$ is the $\mathbb{R}$-vector space of bounded linear operators of V into W) defined by

$$\mu(x) = aa^*xc^*a.$$

μ is injective and satisfies $\mu(xx^*x) = \mu(x)\mu(x)^*\mu(x)$. Assume $\mu(x) = 0$, we have (lemma 9) $0 = aa^*xc^*ac^* = \epsilon aa^*xb^*$ and hence $aa^*xb^*b = 0$. It follows that $xx^* = bb^*xa^*ax^*bb^* + aa^*xb^*bx^*aa^* = 0$ and hence $x = 0$. The injectivity of μ is thus proved. The proof

of $\mu(xx^*x) = \mu(x)\mu(x)^*\mu(x)$ is a trivial computation.
Hence we have proved that $\mu\psi_a\psi_b(j)$ is a $\mathbb{R}$-J*-algebra $\mathbb{R}$-J*-isomorphic to $\psi_a\psi_b(j)$ and hence (lemma 12) it satisfies the hypothesis of lemma 11.
We shall now prove that:

i) $\delta_a(j) \subset \mu\psi_a\psi_b(j)$;

ii) $xy^*x \in \mu\psi_a\psi_b(j)$ whenever $x \in \delta_a(j)$ and $y \in \mu\psi_a\psi_b(j)$.

Let $\xi : \delta_a(j) \to \mu\psi_a\psi_b(j)$ be the linear mapping defined by $\xi(x) = \mu(6\{xcb\})$. Since

$$\xi(x) = aa^*(xc^*b + bc^*x)c^*a = aa^*xc^*bc^*a = \epsilon aa^*xa^*a = \epsilon x,$$

i) follows.
Let $y \in \mu\psi_a\psi_b(j)$ and $h \in \psi_a\psi_b(j)$ be two elements such that $\mu(h) = aa^*hc^*a = y$. We have

$$6\mu(\{xbh\}) = aa^*(xh^*b + bh^*x)c^*a = aa^*xh^*bc^*a = aa^*xa^*ah^*ca^*a =$$

$$= xy^*a$$

and hence ii) is proved.
i) and ii) imply that for $x_1,\ldots,x_n \in \delta_a(j)$

$$x_1a^*\ldots a^*x_n \in \mu\psi_a\psi_b(j).$$

Now let us prove that we have only three possible cases:

I) $\dim \delta_a(j) = 1$ and $\dim \psi_a\psi_b(j) \geqslant 1$;

II) $\dim \delta_a(j) = 2$ and $\dim \psi_a\psi_b(j) = 2$;

III) $\dim \delta_a(j) = 3$ and $\dim \psi_a\psi_b(j) = 4$.

First let us assume that $\dim \delta_a(j) = \dim \psi_a\psi_b(j) = n$. i) implies that $\delta_a(j) = \psi_a\psi_b(j)$, hence we can define in $\delta_a(j)$ a product, *, by $x * y = xa^*y$.
In view of lemma 11, $\delta_a(j)$ turns out to be an associative division algebra

and hence by the Frobenius theorem, isomorphic to $\mathbb{R}$, $\mathbb{C}$ or $\mathbb{H}$. In the above isomorphism the conjugate of $x \in \delta_a(j)$ is $\bar{x} = ax^*a$. Hence this isomorphism is in effect a $\mathbb{R}$-J*-isomorphism since $x * \bar{x} * x = xa^*ax^*aa^*x = xx^*x$, whenever $x \in \delta_a(j)$. Now let us prove that we cannot have $\dim \delta_a(j) = \dim \psi_a\psi_b(j) = 4$. Let us assume that $\dim \delta_a(j) = \dim \psi_a\psi_b(j) = 4$ and let $\{a = a_1, \dots, a_4\}$ be a $\mathbb{R}$-basis of $\delta_a(j)$ (given by lemma 11). Hence the set $\{6\{a_1bc\}, \dots, 6\{a_4bc\}\}$ is a $\mathbb{R}$-basis of $\psi_a\psi_b(j)$ and four real numbers $\lambda_1, \dots, \lambda_4$ exist such that

$$\psi_a\psi_b(j) \ni 6^2\{a_2b\{a_3bc\}\} = a_2a_3^*c + ca_3^*a_2 = \Sigma_i \lambda_i(a_ic^*b + bc^*a_i) =$$

$$= (\epsilon\Sigma_i\lambda_i a_i)a^*c + ca^*(\epsilon\Sigma_i\lambda_i a_i) \quad (i = 1, \dots, 4).$$

Using lemma 9 $(cb^*c = a)$, we obtain $a_2a_3^*a = (\epsilon\Sigma_i\lambda_i a_i)$ and $aa_3^*a_2 = (\epsilon\Sigma_i\lambda_i a_i)$. Since $a_2a_3^*a = -aa_3^*a_2$ (lemma 11), we obtain the contradiction $a_2a_3^*a = 0$.

Now let us prove that $\dim \delta_a(j) = 2 < \dim \psi_a\psi_b(j)$ is impossible. Assume $\{a = a_1, a_2\}$ is a basis of $\delta_a(j)$, given by lemma 11, and $\dim \psi_a\psi_b(j) > 2$. Set

$$6\epsilon\{abc\} = c = c_1, \qquad 6\{a_2bc\} = c_2.$$

It is trivial to prove that c_1 and c_2 are non zero, idempotent elements satisfying

$$c_2 + c_1c_2^*c_1 = (a_2c^*b + bc^*a_2) + c(a_2c^*b + bc^*a_2)^*c = a_2c^*b +$$

$$+ bc^*a_2 + \epsilon ca_2^*a + \epsilon aa_2^*c = a_2c^*b + bc^*a_2 + bc^*aa_2^*a + aa_2^*ac^*b = 0$$

(since $a_2 + aa_2^*a = 0$ by lemma 11). Then, there exists a third non zero, idempotent element $c_3 \in \psi_a\psi_b(j)$ such that $\{c_1, c_2, c_3\}$ is a set that satisfies 1), ..., 5) of lemma 11. Moreover (again by lemma 11) c_3 can be chosen so as to satisfy

$6\{abc_3\} = \eta c_3$ with $\eta \in \{1, -1\}$.

Now $6\{a_1bc_3\} = a_1c_3^*b + bc_3^*a_1 = \eta(a_1a^*c_3 + c_3a^*a_1) = \eta(a_1a^*cc^*c_3 +$

$+ c_3c^*ca^*a_1) = \epsilon\eta(c_2c^*c_3 + c_3c^*c_2) = 0$

gives $a_1a^*c = 0$ and hence the contradiction $a_1a^* = 0$.

Let us conclude this part of the proof showing that $3 \leqslant \dim \delta_a(j) < \dim \psi_a\psi_b(j)$ and $\dim \psi_a\psi_b(j) > 4$ are incompatible.

Let $\{a = a_1, a_2, a_3\} \subset \delta_a(j)$ be a set of non zero, idempotent elements of $\delta_a(j)$ that satisfy 5) of lemma 11. The set $\{a_1, a_2, a_3, a_2a^*a_3\} \subset \mu\psi_a\psi_b(j)$ has non zero idempotent elements that satisfy 5) of lemma 11. Since we assumed $\dim \delta_a(j) = \dim \psi_a\psi_b(j) > 4$, then there exists a non zero, idempotent element $c \in \mu\psi_a\psi_b(j)$ such that

$$ca_i^* + a_ic^* = c^*a_i + a_i^*c = 0 \quad \text{whenever} \quad i = 1,2,3;$$

$$c(a_2a^*a_3)^* + (a_2a^*a_3)c^* = c^*(a_2a^*a_3) + (a_2a^*a_3)^*c = 0.$$

These relations imply $a_2a^*a_3c = 0$ and hence $c = 0$. This contradiction completes the proof that only cases I), II) and III) can occur.

Now let us handle separately these three cases.

CASE I). ($\dim \delta_a(j) = 1$, $\dim \psi_a\psi_b(j) = n \geqslant 1$).

Let $\{c = c_1, \dots, c_n\}$ be a basis of $\psi_a\psi_b(j)$ given by 5) of lemma 11. Let $\epsilon \in \{1, -1\}$ be a coefficient such that $6\{abc\} = \epsilon c$. We prove that $6\{abc_i\} = \epsilon c_i$ whenever $i = 1, \dots, n$. Assume that i is an integer $(1 < i \leqslant n)$ such that $6\{abc_i\} - \epsilon c_i \neq 0$. Then (see lemma 9) there exists a $\lambda \in \mathbb{R}$ such that $x = \lambda(6\{abc_i\} - \epsilon c_i)$ is a non zero, idempotent, irreducible (in $\psi_a\psi_b(j)$) element of $\psi_a\psi_b(j)$. Clearly x satisfies the two relations:

$$6\{abx\} = -\epsilon x,$$

$$c_1x^* + xc_1^* = c_1^*x + x^*c_1 = 0.$$

Hence we have

$$6\{c_1 x b\} = c_1 b^* x + x b^* c_1 = 2\epsilon a c_1^* x.$$

Since $\dim \delta_a(j) = 1$, we must have $2\epsilon a c_1^* x = \tau a$, and $2\epsilon x^* c_1 a^* = -2\epsilon c_1^* x a^* = \tau a^*$. Then $\tau a a^* = 2\epsilon a c_1^* x a^* = -\tau a a^* = 0$, and $a c_1^* x = 0$. Similarly we have $b c^* x = 0$. The above relations imply the contradictior $c_1^* x = 0$ hence $x = 0$.

The set $H = \{a, -\epsilon b\} \cup \{c_1, \dots, c_n\}$ is a basis of j'. The mapping $\sigma . H \to T_{2,n}$ defined by.

$$\sigma(a) = \Omega^n(T^1_1); \quad \sigma(-\epsilon b) = \Omega^n(\tilde{T}^1_1);$$

$$\sigma(c_i) = \Lambda\left(\begin{Vmatrix} 0 & A_i \\ A_i^* & 0 \end{Vmatrix}\right), \ i = 1, \dots, n,$$

is a bijection that can be extended to a J*-isomorphism of j' onto $I_{2,n}$ as a trivial computation can prove (here Ω and Λ are the mappings of lemma 20, T^1_1 and $\tilde{T}^1_1$ the matrices of lemma 22 and A_i the matrices of lemma 21).

CASE II). ($\dim \delta_a(j) = \dim \psi_a \psi_b(j) = 2$).

Let $a = a_1, a_2$ be a basis of $\delta_a(j)$ satisfying 5) of lemma 11 and let c be a non zero, idempotent, irreducible (in $\psi_a \psi_b(j)$) element of $\psi_a \psi_b(j)$ such that $6\{abc\} = \epsilon c$ with $\epsilon \in \{1, -1\}$. Moreover let us set: $c_1 = \epsilon 6\{a_1 bc\}$, $c_2 = \epsilon 6\{a_2 bc\}$, $b_1 = \delta_b(3\{a_1 cc\})$ and $b_2 = -\delta_b(3\{a_2 cc\})$.
The set $H = \{a_1, a_2, b_1, b_2, c_1, c_2\}$ is a basis of j' and the mapping $\sigma : H \to S(2;\mathbb{C})$ defined by:

$$\sigma(a_1) = \begin{Vmatrix} 1 & 0 \\ 0 & 0 \end{Vmatrix}, \quad \sigma(a_2) = \begin{Vmatrix} i & 0 \\ 0 & 0 \end{Vmatrix},$$

$$\sigma(b_1) = \begin{Vmatrix} 0 & 0 \\ 0 & 1 \end{Vmatrix}, \quad \sigma(b_2) = \begin{Vmatrix} 0 & 0 \\ 0 & i \end{Vmatrix},$$

$$\sigma(c_1) = \begin{Vmatrix} 0 & 1 \\ 1 & 0 \end{Vmatrix}, \quad \sigma(c_2) = \begin{Vmatrix} 0 & i \\ i & 0 \end{Vmatrix},$$

is a bijection of H onto a basis of $S(2;\mathbb{C})$. Moreover σ preserves the real structure formulae. We conclude (see lemma 18) that $\mathcal{J}'$ and $S(2;\mathbb{C})$ are $\mathbb{R}$-J*-isomorphic.

CASE III). ($\dim \delta_a(\mathcal{J}) = 3$, $\dim \psi_a\psi_b(\mathcal{J}) = 4$).

Let $\{a = a_1, a_2, a_3\}$ be a basis of $\delta_a(\mathcal{J})$ which satisfies 5) of lemma 11, and c a non zero, idempotent, irreducible (in $\psi_a\psi_b(\mathcal{J})$) element of $\psi_a\psi_b(\mathcal{J})$ satisfying $6\{abc\} = \epsilon c$ with $\epsilon \in \{1, -1\}$.

Let us define:

$$c_0 = \epsilon 6\{a_2 b c_3\}, \quad c_i = \epsilon 6\{a_i bc\} \quad i = 1,2,3;$$

$$b_i = \epsilon \delta_b(3\{a_i cc\}) \quad i = 1,2,3.$$

The set $H = \{a_1, a_2, a_3, c_0, c_1, c_2, c_3, b_1, b_2, b_3\}$ is a basis of $\mathcal{J}'$ and the mapping $\sigma : H \to H^*(2;\mathbb{H})$ defined by:

$$\sigma(a_1) = \begin{Vmatrix} i & 0 \\ 0 & 0 \end{Vmatrix}, \quad \sigma(a_2) = \begin{Vmatrix} j & 0 \\ 0 & 0 \end{Vmatrix}, \quad \sigma(a_3) = \begin{Vmatrix} k & 0 \\ 0 & 0 \end{Vmatrix},$$

$$\sigma(b_1) = \begin{Vmatrix} 0 & 0 \\ 0 & i \end{Vmatrix}, \quad \sigma(b_2) = \begin{Vmatrix} 0 & 0 \\ 0 & j \end{Vmatrix}, \quad \sigma(b_3) = \begin{Vmatrix} 0 & 0 \\ 0 & k \end{Vmatrix},$$

$$\sigma(c_0) = \begin{Vmatrix} 0 & 1 \\ -1 & 0 \end{Vmatrix}, \quad \sigma(c_1) = \begin{Vmatrix} 0 & i \\ i & 0 \end{Vmatrix}, \quad \sigma(c_2) = \begin{Vmatrix} 0 & j \\ j & 0 \end{Vmatrix},$$

$$\sigma(c_3) = \begin{Vmatrix} 0 & k \\ k & 0 \end{Vmatrix},$$

maps H onto a basis of $H^*(2;\mathbb{H})$ and can be extended to a $\mathbb{R}$-J*-isomorphism of $\mathcal{J}'$ onto $H^*(2;\mathbb{H})$.

The proof is now complete.

<u>LEMMA 29</u>. Let a and b be two non zero, idempotent irreducible, strongly independent elements of a $\mathbb{R}$-J*-algebra j, such that $\psi_a\psi_b(j) \neq \{0\}$. $\psi_a\psi_b(j)$ has a subset B such that:

1) any element of B is non zero, idempotent and irreducible in j;

2) if $x \in B$ then $\tilde{x} = 6\{abx\}$ is <u>not</u> in B;

3) if $x \in B \cup \tilde{B} \cup \{a,b\}$ then x and $\tilde{x}$ are strongly independent; (set $\tilde{B} = \{\tilde{x} : x \in B\}$, $\tilde{\tilde{x}} = x$, $\tilde{a} = b$ and $\tilde{b} = a$);

4) if $x, y \in B \cup \tilde{B} \cup \{a,b\}$, where $x,y,\tilde{x},\tilde{y}$ are distinct elements, then $y = xx^*y + yx^*x$;

5) if $x, y \in B \cup \tilde{B} \cup \{a,b\}$, where $x,y,\tilde{x},\tilde{y}$ are distinct elements, then $\delta_x(y) = xx^*yx^*x = 0$;

6) if $x, y \in B \cup \tilde{B} \cup \{a,b\}$, with $x,y,\tilde{x},\tilde{y}$ distinct elements, then $6\{x\tilde{x}y\} = \tilde{y}$;

7) for $x, y \in B \cup \tilde{B} \cup \{a,b\}$, where $x,y,\tilde{x},\tilde{y}$ are distinct elements, and for $z \in \psi_x\psi_{\tilde{x}}\psi_y\psi_{\tilde{y}}(j)$, we have $\{x\tilde{x}\{y\tilde{y}z\}\} = \{y\tilde{y}\{x\tilde{x}z\}\}$;

8) if $x, y, z \in B \cup \tilde{B} \cup \{a,b\}$, where $x,y,z,\tilde{x},\tilde{y},\tilde{z}$ are distinct elements, then $\{xyz\} = 0$;

9) $\psi_a\psi_b(j) = \Sigma_x\ \delta_x(j) + N(j)$ $(x \in B \cup \tilde{B})$ where $N(j)$ is a $\mathbb{R}$-J*-subalgebra of j which does <u>not</u> contain any non zero, idempotent irreducible (in j) element.

<u>PROOF</u>. Let us construct our set B by using a finite induction argument.

If $\psi_a\psi_b(j)$ does not have any non zero, idempotent element which is irreducible in j, our lemma is trivially verified with $B = \{\emptyset\}$.

Assume that B_r is a set of r elements, $r \geqslant 0$, satisfying 1), ..., 8) and such that there exists a non zero, idempotent element $c \in N_r(j) = (\pi_x\psi_x)(j)$ $(x \in B_r \cup \tilde{B}_r)$, such that c is irreducible in j. Whenever $x \in B_r \cup \{a\}$ and $y \in N_r(j)$, we have $6\{x\tilde{x}y\} \in N_r(j)$. In fact, from

$$\psi_x(6\{x\tilde{x}y\}) = xx^*(xy^*\tilde{x} + \tilde{x}y^*x) + (xy^*\tilde{x} + \tilde{x}y^*x)x^*x = xx^*(xy^*\tilde{x} + \tilde{x}y^*x)x^*x =$$
$$= xy^*\tilde{x} + xy^*\tilde{x}$$

we obtain that $6\{x\tilde{x}y\} \in \psi_x(j)$. Analogously, we can prove that $6\{x\tilde{x}y\} \in \psi_{\tilde{x}}(j)$. Suppose now $z \in B_r \cup \tilde{B}_r \cup \{a,b\}$ is an element such that $z, x, \tilde{z}, \tilde{x}$ are distinct. Then

$$\psi_z(6\{x\tilde{x}y\}) = zz^*(xy^*\tilde{x} + \tilde{x}y^*x) + (xy^*\tilde{x} + \tilde{x}y^*x)z^*z - zz^*(xy^*\tilde{x} + \tilde{x}y^*x)z^*z =$$

$$= xy^*\tilde{x} + xy^*\tilde{x}$$

(see hypothesis 4) gives

$$6\{x\tilde{x}y\} \in \cap_z \psi_z(j) = N_r(j) \quad (z \in B_r \cup B_r \cup \{a,b\}).$$

Whenever $x \in B_r \cup \{a\}$, let us denote $\omega_x \cdot N_r(j) \to N_r(j)$ the linear mapping defined by $\omega_x(y) = 6\{x\tilde{x}y\}$. Moreover let us consider the set

$$I = \{\pi_w \omega_w(c) : w \in S \subset B_r \cup \{a\} \text{ where } |S| \text{ is even}\}.$$

We can prove that I is a finite subset of $\delta_c(j)$. I is a finite set since $B_r \cup \{a\}$ is. Assume now $h \in \delta_c(j)$ and $x, y \in B_r \cup \{a\}$ with $x \neq y$. We have

$$\delta_c(\omega_x \omega_y(h)) = \omega_x \omega_y \delta_c(h) = \omega_x \omega_y(h)$$

since $c \in \psi_x \psi_{\tilde{x}} \psi_y \psi_{\tilde{y}}(j)$. A trivial induction argument now completes the proof.

Let $t = \Sigma_x x$ $(x \in I)$. $t \in \delta_c(j)$. Let us prove that $t \neq 0$. Assume $t = \Sigma_S(\pi_w \omega_w)(c) = 0$ $(w \in S \subset B_r \cup \{a\}$ with $|S|$ even). Hence, if $p \in B_r \cup \{a\}$, we have

$$0 = p^*(\Sigma_S(\pi_w \omega_w)(c))\tilde{p}^* = p^*p(\Sigma_R(\pi_u \omega_u)(c))\widetilde{pp}^* + p^*(\Sigma_K(\pi_y \omega_y)(c))\tilde{p}^*$$

$(w \in S \subset B_r \cup \{a\}$ with $|S|$ even; $u \in R \cup (B_r \cup \{a\}) - \{p\}$ with $|R|$ odd; $y \in K \cup (B_r \cup \{a\}) - \{p\}$ with $|K|$ even).
It follows that (see hypothesis 4) $\Sigma_S(\pi_w \omega_w)(c) = 0$
$(w \in S \subset (B_r \cup \{a\}) - \{p\})$ but this implies, by a trivial induction

argument, $c = 0$, a contradiction.
Since $\delta_c(j)$ is a J^*-algebra which satisfies the assumptions of lemma 11, t is irreducible (in $\delta_c(j)$) and $tt^*t \doteq t$. Let $\lambda \in \mathbb{R}$ be a coefficient such that λt is a non zero, idempotent element. Whenever $h \in j$, we have:

$$\lambda^4 tt^*ht^*t + \lambda^2 th^*t = \lambda^4 tt^*\delta_c(h)t^*t + \lambda^2 t\delta_c(h)^*t \doteq t$$

and λt is irreducible in j. Set $p = \lambda t$ and $\tilde{p} = 6\{abp\}$. We can prove that $B_{r+1} = B_r \cup \{p\}$ satisfies 1), ..., 8). 1) follows from the definition of p. 2), ..., 5) follow from lemma 13.
If $x \in B_r \cup \{a\}$ we have

$$6\{x\tilde{x}p\} = \lambda\omega_x(\Sigma_S(\pi_w\omega_w)(c)) = \lambda\Sigma_M(\pi_u\omega_u)(c) + \lambda\Sigma_H(\pi_h\omega_h)(c) =$$

$$= \lambda\Sigma_N(\pi_n\omega_n)(c) = \tilde{p}$$

($w \in S \subset B_r \cup \{a\}$ with $|S|$ even; $u \in M \subset (B_r \cup \{a\}) - \{x\}$ with $|M|$ even; $h \in H \subset (B_r \cup \{a\}) - \{x\}$ with $|H|$ odd; $n \in N \subset B_r \cup \{a\}$ with $|N|$ odd). $6\{\widetilde{x\tilde{x}p}\} = p$ can be proved in a similar way. On the other hand we have

$$6\{p\tilde{p}x\} = px^*\tilde{p} + \tilde{p}x^*p = px^*xp^*\tilde{x} + \tilde{x}p^*xx^*p = pp^*\tilde{x} + \tilde{x}p^*p = \tilde{x}.$$

$6\{\widetilde{p\tilde{p}x}\} = x$ can be proved similarly. 6) is thus proved.
Whenever $x \in B_r \cup \{a\}$ and $y \in (\pi_z\psi_z)(j)$ (where $z \in B_{r+1} \cup \tilde{B}_{r+1} \cup \{a,b\}$) we have (see 4) and 6) just proved above).

$$\omega_p\omega_x(y) = px^*y\tilde{x}^*\tilde{p} + \tilde{p}x^*y\tilde{x}^*p + p\tilde{x}^*yx^*p + \widetilde{\widetilde{px}}^*yx^*p = \widetilde{\widetilde{xx}}^*px^*\tilde{y}x^*\tilde{p}x^*x +$$

$$+ \widetilde{\widetilde{xx}}^*\tilde{p}x^*y\tilde{x}^*px^*x + xx^*p\tilde{x}^*yx^*\widetilde{\widetilde{px}}^*\tilde{x} + xx^*\widetilde{\widetilde{px}}^*yx^*p\tilde{x}^*\tilde{x} =$$

$$= x\widetilde{\widetilde{p}}^*yp^*x + \tilde{x}p^*\tilde{y}p^*x + x\tilde{p}^*yp^*\tilde{x} + xp^*y\tilde{p}^*x = \omega_x\omega_p(y)$$

hence 7) is proved.

Now let $x, y \in \mathcal{B}_r \cup \widetilde{\mathcal{B}}_r \cup \{a,b\}$, with $x, y, \tilde{x}, \tilde{y}$ distinct elements. 6) gives $6\{xyp\} = 6\{xy\tilde{p}\} = 0$ and this proves 8).

Lemma 13 implies

$$N_r(j) = \delta_p(N_r(j)) + \delta_{\tilde{p}}(N_r(j)) + \psi_p\psi_{\tilde{p}}(N_r(j)).$$

Hence the above induction argument can be carried on until 9) is verified.

The proof is now complete.

<u>LEMMA 30</u>. Let a and b be two non zero, idempotent, irreducible, strongly independent elements of a J*-algebra j such that $\psi_a\psi_b(j) \neq 0$. Moreover let $\mathcal{B}_{ab}$ be a subset of $\psi_a\psi_b(j)$ which satisfies 1), ..., 9) of lemma 29.

Then we have:

1) for $x, y \in \mathcal{B}_{ab} \cup \widetilde{\mathcal{B}}_{ab} \cup \{a,b\}$, $\delta_x(j)$ and $\delta_y(j)$ are isomorphic;

2) if $|\mathcal{B}_{ab}| = 1$, then $\dim \delta_a(j) = 1,2,4$;

3) if $|\mathcal{B}_{ab}| > 1$, then $\dim \delta_a(j) = 1,2$.

<u>PROOF</u>. Lemma 9 implies $\delta_a(j)$ and $\delta_b(j)$ are isomorphic, hence 1) is proved if $\mathcal{B}_{ab} = \emptyset$. Assume $\mathcal{B}_{ab} \neq \emptyset$. Using the notation established in lemma 29, we have, for $x, y \in \mathcal{B}_{ab} \cup \widetilde{\mathcal{B}}_{ab} \cup \{a,b\}$, with $x, y, \tilde{x}, \tilde{y}$ distinct elements, and $z \in \delta_x(j)$,

$$\delta_y(j) \ni \delta_y(6\,\widetilde{\{xyz\}}) = yy^*(\widetilde{xy}^*z + x\tilde{y}^*\tilde{x})y^*y = \widetilde{xy}^*z + z\tilde{y}^*\tilde{x} = 6\,\widetilde{\{zxy\}}.$$

Hence we can define a linear mapping $\varphi_{xy} : \delta_x(j) \to \delta_y(j)$, by $\varphi_{xy}(z) = 6\,\widetilde{\{xyz\}}$. $\varphi_{xy}\varphi_{yx}$ is the identity on $\delta_x(j)$. In fact, whenever $z \in \delta_x(j)$, we have

$$\varphi_{xy}\varphi_{yx}(z) = (\widetilde{xy}^*z + z\tilde{y}^*x)\tilde{x}^*\tilde{y} + \widetilde{yx}^*(\widetilde{xy}^*z + z\tilde{y}^*x) = z\tilde{y}^*\widetilde{xx}^*\tilde{y} + \widetilde{yx}^*\widetilde{xy}^*z = z.$$

1) follows trivially.

Now let us prove that we can choose a $\mathbb{R}$-basis of $\delta_a(j)$, $\{a = a_1,\dots,a_n\}$ that satisfies 5) of lemma 11 and

$$(1)\quad a_i = \omega_x\omega_y(a_i),\quad i = 1,\dots,n \text{ and } x,\, y \in \mathcal{B}_{ab}.$$

We shall obtain $\{a_1,\dots,a_n\}$ using an induction argument. Let x,y in $\mathcal{B}_{ab}$. Since $\tilde{x} = 6\{abx\}$ and $\tilde{y} = 6\{aby\}$, using 4) and 6) of lemma 29, we have

$$\omega_x\omega_y(a) = xy^*a\tilde{y}^*\tilde{x} + x\tilde{y}^*ay^*\tilde{x} + \tilde{x}y^*a\tilde{y}^*x + \tilde{x}\tilde{y}^*ay^*x = xy^*yx^*a +$$

$$+ x\tilde{y}^*\tilde{y}x^*a + x\tilde{y}^*y\tilde{x}^*a + \tilde{x}\tilde{y}^*\tilde{y}\tilde{x}^*a = a.$$

Assume now that $a = a_1, a_2, \dots a_r$, $r < m$, be r non zero, idempotent elements satisfying 5) of lemma 11 and (1). Whenever $h \in \delta_a(j)$ is such that $h + a_p h^* a_p = 0$, $p = 1,\dots,r$, we have

$$\omega_x\omega_y(h) + a_p\omega_x\omega_y(h)^*a_p = \omega_x\omega_y(h) + \omega_x\omega_y(a_p)\omega_x\omega_y(h)^*\omega_x\omega_y(a_p) =$$

$$= \omega_x\omega_y(h + a_p h^* a_p) = 0.$$

Thus if $h \neq 0$ (see the proof of lemma 29) it follows that $t = \Sigma_S(\pi_w\omega_w)(h) \neq 0$ ($w \in S \subset \mathcal{B}_{ab} \cup \tilde{\mathcal{B}}_{ab} \cup \{a,b\}$ with $|S|$ even) and $t + a_p t^* a_p = 0$ whenever $1 \leqslant p \leqslant r$.
Now we define a_{p+1} to be a non zero, idempotent element such that $a_{p+1} \doteq t$. The induction argument is complete.

Assume that $\dim \delta_a(j) \geqslant 3$ and $|\mathcal{B}_{ab}| \geqslant 1$. Moreover let $\{a = a_1, a_2, a_3\}$ be a set of non zero idempotent elements of $\delta_a(j)$ satisfying 5) of lemma 11 and (1) above. Whenever $x \in \mathcal{B}_{ab}$, we have

$$\delta_a(j) \ni 6^3\{b\{xba_2\}\{\tilde{x}ba_3\} = (a_3\tilde{x}^*b + b\tilde{x}^*a_3)\, b^*(a_2x^*b + bx^*a_2) +$$

$$+ (a_2x^*b + bx^*a_2)b^*(a_3\tilde{x}^*b + b\tilde{x}^*a_3) = a_3\tilde{x}^*bx^*a_2 + a_2x^*b\tilde{x}^*a_3 =$$

$$= a_3a_1^*a_2\tilde{x}^*\tilde{x} + a_2a_1^*a_3\tilde{x}^*\tilde{x} - a_1a_2^*a_3x^*x = a_1a_2^*a_3x^*x = k.$$

Our assumption implies that $k \neq 0$. In fact, assume $k = 0$; then we have $a_1a_2^*a_3x^* = 0$ and $a_1a_2^*a_3\tilde{x}^* = 0$. Hence $a_1a_2^*a_3 = 0$ and this is a contradiction.

Now it is trivial to prove that $ka_i^* + a_ik^* = k^*a_i + a_i^*k = 0$ whenever $i = 1,2,3$. Hence, by lemma 11, $\dim \delta(j) \geq 4$.

Let us prove that we cannot have $\dim \delta_a(j) \geq 4$ and $|B_{ab}| \geq 2$. Assume the above condition is false and let $x, y \in B_{ab}$, with $x \neq y$. Moreover let $\{a = a_1, a_2, a_3\}$ be the set defined above. We must have:

$$\delta_a(j) \ni 6^3\{b\{xba_2\}\{yba_3\} = a_3y^*bx^*a_2 + a_2x^*by^*a_3 = a_3a_1^*a_2y^*\tilde{x} -$$

$$- a_2a_1^*a_3x^*\tilde{y} = 2a_1a_2^*a_3x^*\tilde{y};$$

and (obtained similarly)

$$\delta_a(j) \ni 2a_1a_2^*a_3x^*y,\ 2a_1a_2^*a_3\tilde{x}^*\tilde{y},\ 2a_1a_2^*a_3\tilde{x}^*y.$$

Almost one of these four elements of $\delta_a(j)$ must be non zero, otherwise we would have the contradiction $a_1a_2^*a_3 = 0$. Assume $a_1a_2^*a_3x^*\tilde{y} = a_1a_2^*a_3y^*\tilde{x} = t \neq 0$. Since $a_it^* + ta_i^* = t^*a_i + a_i^*t = 0$ whenever $i = 1,2,3$, a_4 can be chosen to be a non zero, idempotent element proportional to t. Hence we have $a_4y^* = a_4x^* = 0$ and (see lemma 11)

$$a_iy^* = a_4a_4^*a_ia_4^*a_4y^* = 0;\ a_ix^* = 0$$

whenever $i = 1,2,3,4$. Since $a = ax^*x + a\tilde{x}^*x^* = ay^*y + a\tilde{y}^*\tilde{y}$ (see lemma 29), we have $a = a\tilde{x}^*\tilde{x} = a\tilde{y}^*\tilde{y}$ and hence $a^*a = a^*a\tilde{x}^*\tilde{x} = \tilde{x}^*\tilde{x} - \tilde{x}^*aa^*\tilde{x} = \tilde{x}^*\tilde{x} - \tilde{x}^*a\tilde{x}^*\tilde{x}a^*\tilde{x} = \tilde{x}^*\tilde{x}$. It follows that $\tilde{x}^*\tilde{x} = \tilde{x}^*\widetilde{xy^*}\tilde{y}$ and $\widetilde{xy^*}\tilde{y} = \tilde{x}$, hence $x\tilde{y}^*\tilde{x} = \widetilde{yy^*}\tilde{x} = 0$ and $\tilde{y}^*\tilde{x} = 0$. Similarly we obtain $\widetilde{yx^*} = 0$ but we have a contradiction since x and y are not strongly independent.

To conclude the proof we have only to show that it is impossible to have $\dim \delta_a(j) > 4$. Assume that $\dim \delta_a(j) > 4$ and let $\{a = a_1, \dots, a_5\}$ be a subset of $\delta_a(j)$ that satisfies 5) of lemma 11,

(1), and $a_4 \doteq a_1 a_2^* a_3 \tilde{x}^* \tilde{x} - a_1 a_2^* a_3 x^* x$ (where x is an element of $\mathcal{B}_{ab}$). We have

$$0 = a_5 a_4^* + a_4 a_5^* = a_5 \tilde{x}^* \tilde{x} a_3^* a_2 a_1^* - a_5 x^* x a_3^* a_2 a_1^* + a_4 a_5^* = 2a_4 a_5^*$$

and this is a contradiction.

The proof is complete.

Using the notation already established, we have the following lemma:

LEMMA 31. Let a and b be two non zero, idempotent, irreducible, strongly independent elements of $\mathbb{R}$-J*-algebra $\mathcal{J}$ such that $\psi_a \psi_b(\mathcal{J}) \neq \{0\}$. Let $\mathcal{B}_{ab}$ be a subset of $\psi_a \psi_b(\mathcal{J})$ whose existence is given by lemma 29, and assume $\mathcal{N}(\mathcal{J}) = \cap_x \psi_x(\mathcal{J}) = 0$ $(x \in \mathcal{B}_{ab} \cup \tilde{\mathcal{B}}_{ab} \cup \{a,b\})$. Then the $\mathbb{R}$-J*-algebra

$$\mathcal{J}' = \delta_a(\mathcal{J}) + \delta_b(\mathcal{J}) + \psi_a \psi_b(\mathcal{J}) = (\Sigma_x \delta_x)(\mathcal{J}) \quad (x \in \mathcal{B}_{ab} \cup \tilde{\mathcal{B}}_{ab} \cup \{a,b\})$$

is $\mathbb{R}$-J*-isomorphic to one of the following

$$\mathcal{M}(2;\mathbb{H}), \quad \mathcal{I}_{2n}, \quad \mathcal{I}_{2n}^{\mathbb{C}}.$$

PROOF. By lemma 30, we limit our proof to the following three cases:

I) $|\mathcal{B}_{ab}| = 1$ and $\dim \delta_a(\mathcal{J}) = 4$

II) $|\mathcal{B}_{ab}| \geqslant 1$ and $\dim \delta_a(\mathcal{J}) = 1$

III) $|\mathcal{B}_{ab}| \geqslant 1$ and $\dim \delta_a(\mathcal{J}) = 2$.

CASE I. $(|\mathcal{B}_{ab}| = 1$ and $\dim \delta_a(\mathcal{J}) = 4)$.

Let $\mathcal{B}_{ab} = \{x\}$ and $\{a = a_1, a_2, a_3, a_4\}$ be a $\mathbb{R}$-basis of $\delta_a(\mathcal{J})$ which satisfies 5) of lemma 11 and $a_i = \omega_a \omega_x(a_i)$ whenever $i = 1,2,3,4$ (See the proof of lemma 30 for the existence of such a basis). It is trivial to prove that the sets:

$$\{x_i = \varphi_{xa}(a_i) : i = 1,2,3,4\},$$

$$\{\tilde{x}_i = \varphi_{\tilde{x}a}(a_i) : i = 1,2,3,4\},$$

$$\{b_i = \varphi_{bx}\varphi_{xa}(a_i) : i = 1,2,3,4\},$$

($\varphi_{xa}, \varphi_{\tilde{x}a}, \varphi_{bx}$ are the mapping defined in the proof of lemma 30) are $\mathbb{R}$-bases of $\delta_x(j)$, $\delta_{\tilde{x}}(j)$ and $\delta_b(j)$ respectively satisfying 5) of lemma 11. We have

$$\delta_{\tilde{x}}(j) \ni 6\{a_2 b_4 x_3\} = a_2 x_3^* b_4 + b_4 x_3^* a_2 = a_2 a_3^* \tilde{x} b^* b \tilde{x}^* a_4 a^* \tilde{x} +$$

$$+ \tilde{x} a^* a_4 \tilde{x}^* b b^* \tilde{x} a_3^* a_2 = - a a_2^* a_3 a_4^* \tilde{x} - b b_2^* b_3 b_4^* \tilde{x} = t.$$

t is different from zero. In fact, assume $t = 0$. Then it follows $a_4^* \tilde{x} = b_4^* \tilde{x} = 0$ and hence the contradiction $\tilde{x} = 0$. Now, it is easy to prove that t is idempotent and satisfies. $\tilde{x}_i^* t + t^* \tilde{x}_i = \tilde{x}_i t^* + t \tilde{x}_i^* = 0$ whenever $i = 1,\dots,4$.
Hence $t = -\epsilon \tilde{x}$ with $\epsilon \in \{1, -1\}$.

Let σ be the mapping of $\{a_i, b_i, x_i, \tilde{x}_i : i = 1,\dots,4\}$ into $M(2;\mathbb{H})$ defined by:

$$\sigma(a_1) = \begin{Vmatrix} 1 & 0 \\ 0 & 0 \end{Vmatrix}, \sigma(a_2) = \begin{Vmatrix} i & 0 \\ 0 & 0 \end{Vmatrix}, \sigma(a_3) = \begin{Vmatrix} j & 0 \\ 0 & 0 \end{Vmatrix}, \sigma(a_4) = \epsilon \begin{Vmatrix} k & 0 \\ 0 & 0 \end{Vmatrix},$$

$$\sigma(b_1) = \begin{Vmatrix} 0 & 0 \\ 0 & 1 \end{Vmatrix}, \sigma(b_2) = \begin{Vmatrix} 0 & 0 \\ 0 & i \end{Vmatrix}, \sigma(b_3) = \begin{Vmatrix} 0 & 0 \\ 0 & j \end{Vmatrix}, \sigma(b_4) = \epsilon \begin{Vmatrix} 0 & 0 \\ 0 & k \end{Vmatrix},$$

$$\sigma(x_1) = \begin{Vmatrix} 0 & 1 \\ 0 & 0 \end{Vmatrix}, \sigma(x_2) = \begin{Vmatrix} 0 & i \\ 0 & 0 \end{Vmatrix}, \sigma(x_3) = \begin{Vmatrix} 0 & j \\ 0 & 0 \end{Vmatrix}, \sigma(x_4) = \epsilon \begin{Vmatrix} 0 & k \\ 0 & 0 \end{Vmatrix},$$

$$\sigma(x_1) = \begin{Vmatrix} 0 & 0 \\ 1 & 0 \end{Vmatrix}, \sigma(x_2) = \begin{Vmatrix} 0 & 0 \\ i & 0 \end{Vmatrix}, \sigma(x_3) = \begin{Vmatrix} 0 & 0 \\ j & 0 \end{Vmatrix}, \sigma(x_4) = \epsilon \begin{Vmatrix} 0 & 0 \\ k & 0 \end{Vmatrix},$$

Now it is trivial to prove, following lemma 18, that σ extends to a $\mathbb{R}$-J*-isomorphism of j' onto $M(2;\mathbb{H})$.

CASE II. ($|\mathcal{B}_{ab}| = n \geq 1$ and $\dim \delta_a(j) = 1$).

Let $\sigma : \mathcal{B}_{ab} \cup \tilde{\mathcal{B}}_{ab} \cup \{a,b\} \to T_{2n+2}$ be a bijective mapping such that $\sigma(\tilde{x}) = \widetilde{\sigma(x)}$ whenever $x \in \mathcal{B}_{ab} \cup \tilde{\mathcal{B}}_{ab} \cup \{a,b\}$. It follows from lemma 18 and lemma 30, that σ extends to a $\mathbb{R}$-J*-isomorphism of j' onto I_{2n+2}.

CASE III. ($|\mathcal{B}_{ab}| = n \geq 1$ and $\dim \delta_a(j) = 2$).

Let $\{a = a_1, a_2\}$ be a basis of $\delta_a(j)$ which satisfies 5) of lemma 11 and such that $\omega_x\omega_y(a_i) = a_i$ whenever $i = 1,2$ and $x, y \in \mathcal{B}_{ab} \cup \{a\}$.

Let $\sigma : \mathcal{B}_{ab} \cup \{a\} \to \{T^{n+1}_1, \ldots, T^{n+1}_{n+1}\} \subset I_{2n+2}^{\mathbb{C}}$ be an arbitrary bijective mapping. σ extends to a $\mathbb{R}$-J*-isomorphism of j' onto $I_{2n+2}^{\mathbb{C}}$ such that

$$\sigma(\varphi_{xa}(a_1)) = \sigma(x), \quad \sigma(\varphi_{\tilde{x}a}(a_1)) = \widetilde{\sigma(x)}, \quad \sigma(\varphi_{xa}(a_2)) = i\sigma(x),$$

$$\sigma(\varphi_{\tilde{x}a}(a_2)) = \widetilde{i\sigma(x)} \quad \text{whenever} \quad x \in \mathcal{B}_{ab};$$

$$\sigma(a_2) = i\sigma(a), \quad \sigma(\varphi_{bx}\varphi_{xa}(a_1)) = \widetilde{\sigma(a)}, \quad \sigma(\varphi_{bx}\varphi_{xa}(a_2)) = \widetilde{i\sigma(a)}$$

(observe that $\varphi_{bx}\varphi_{xa} : \delta_a(j) \to \delta_b(j)$ is independent of the element $x \in \mathcal{B}_{ab}$).

The proof is complete.

<u>LEMMA 32</u>. Let a and b be two non zero, idempotent, strongly independent elements of an $\mathbb{R}$-J*-algebra j, such that $\psi_a\psi_b(j) \neq \{0\}$. Let $\mathcal{B}_{ab}$ be a subset of $\psi_a\psi_b(j)$ defined in lemma 29. Moreover, assume that

$$N(j) = \cap_x \psi_x(j) \neq \{0\} \quad (x \in \mathcal{B}_{ab} \cup \tilde{\mathcal{B}}_{ab} \cup \{a,b\}).$$

Then the $\mathbb{R}$-J*-algebra

$$j' = \delta_a(j) + \delta_b(j) + \psi_a\psi_b(j)$$

is $\mathbb{R}$-J*-isomorphic to one of the following $\mathbb{R}$-J*-algebras.

$$I_{2n,m}, \quad I_{2n,1}\mathbb{C} \qquad (n \geq 1 \text{ and } m \geq 1).$$

PROOF. By lemmas 28, 29 and 30, we only need to consider the following two cases:

I) $|B_{ab}| = n - 1 \geq 1$, $\dim \delta_a(j) = 1$ and $\dim N(j) \geq 1$;

II) $|B_{ab}| = n - 1 \geq 1$, $\dim \delta_a(j) = 2$ and $\dim N(j) = 2$.

CASE I. ($|B_{ab}| = n - 1 \geq 1$, $\dim \delta_a(j) = 1$ and $\dim N(j) \geq 1$). By lemma 12, $N(j)$ is a J*-algebra which satisfies the assumption of lemma 11. Let c be a non zero, idempotent, irreducible element of $N(j)$. One of the following two elements of $N(j)$.:

$$c_1 = \Sigma_S(\pi_w\omega_w)(c) \qquad (w \in S \subset B_{ab} \cup \{a\}),$$

$$c_2 = \Sigma_S(-1)^{|S|}(\pi_w\omega_w)(c) \qquad (w \in S \subset B_{ab} \cup \{a\})$$

must be non zero (the proof is performed as the proof of lemma 30). Assume $c_1 \neq 0$. Hence $6\{x\tilde{x}c_1\} = c_1$ whenever $x \in B_{ab} \cup \{a\}$. (If we assume $c_2 \neq 0$, we have $6\{x\tilde{x}c_2\} = -c_2$ whenever $x \in B_{ab} \cup \{a\}$). Now we can prove, as in lemma 31, that $N(j)$ has a basis $\{x_1,\dots,x_m\}$ whose elements are non zero and idempotent; moreover they satisfy 5) of lemma 11 and $6\{x\tilde{x}x_i\} = \mu_i x_i$, with $\mu_i \in \{1, -1\}$, whenever $x \in B_{ab} \cup \{a\}$. Assume $m \geq 2$ and let $1 \leq i < j \leq m$. We have $6\{ax_ix_j\} = x_ia^*x_j + x_ja^*x_i = (\mu_i - \mu_j)bx^*_ix_j \doteq b$. Hence $(\mu_i - \mu_j)bx^*_ix_j = (\mu_j - \mu_i)bx^*_jx_i = \sigma b$ and $(\mu_i - \mu_j)bx^*_i = \sigma bx^*_j$, $-(\mu_i - \mu_j)bx^*_j = \sigma bx^*_i$. It follows that $(\mu_i - \mu_j)^2 bx^*_i = -\sigma^2 bx^*_i$ and, since $bx^*_i \neq 0$, $\mu_i - \mu_j = 0$. We have proved that $\mu_1 = \dots = \mu_m = \mu$.

The set $H = B_{ab} \cup \tilde{B}_{ab} \quad \{a,b\} \cup \{x_1,\dots,x_m\}$ is a basis of j'. Now, set two arbitrary bijective mappings $g . \{x_1,\dots,x_m\} \to U_m$ and $f : B_{ab} \cup \{a\} \to T_n$ (U_m and T_n were defined in part 2). The reader can prove that the mapping $\sigma : H \to I_{2n,m}$ defined by.

$$\sigma(x) = \Omega^m f(x) \quad \text{whenever} \quad x \in B_{ab} \cup \{a\};$$

$$\sigma(\tilde{x}) = \Omega^m \widetilde{f(x)} \quad \text{whenever} \quad x \in B_{ab} \cup \{a\};$$

$$\sigma(x_i) = \Lambda^n g(x_i) \quad \text{whenever} \quad i = 1,2,\ldots,m;$$

extends to a $\mathbb{R}$-J*-isomorphism of j' onto $I_{2n,m}$.

CASE II. ($|B_{ab}| = n - 1 \geqslant 1$, $\dim \delta_a(j) = 2$ and $\dim N(j) = 2$). Let $c \in N(j)$ be a non zero, idempotent element such that $6\{x\tilde{x}c\} = \epsilon c$, with $\epsilon \in \{1, -1\}$, whenever $x \in B_{ab} \cup \{a\}$. Moreover let $\{a = a_1, a_2\}$ be a $\mathbb{R}$-basis of $\delta_a(j)$ whose elements are non zero, irreducible and satisfy 5) of lemma 11 and $\omega_x \omega_y(a_i) = a_i$, $i = 1,2$, whenever $x, y \in B_{ab} \cup \{a\}$.

Let $f : B_{ab} \cup \{a\} \to \{T^n_1, \ldots T^n_n\}$ be an arbitrary bijective mapping. The mapping $\sigma : B_{ab} \cup \tilde{B}_{ab} \cup \{a,b,c\} \to I^{\mathbb{C}}_{2n,1}$ is defined by:

$$\sigma(\varphi_{xa}(a_1)) = \Omega f(x), \quad \sigma(\varphi_{xa}(a_2)) = i\Omega f(x), \quad \sigma(\varphi_{\tilde{x}a}(a_1)) = \Omega \widetilde{f(x)},$$

$$\sigma(\varphi_{\tilde{x}a}(a_2)) = i\Omega \widetilde{f(x)}, \quad \sigma(a_1) = \Omega f(a), \quad \sigma(a_2) = i\Omega f(a),$$

$$\sigma(\varphi_{bz}\varphi_{za}(a_1)) = \Omega \widetilde{f(a)}, \quad \sigma(\varphi_{bz}\varphi_{za}(a_2)) = i\Omega \widetilde{f(a)},$$

$$\sigma(c) = i\epsilon \Lambda^n \left(\left\| \begin{matrix} 0 & 1 \\ 1 & 0 \end{matrix} \right\| \right), \quad \sigma(6\{a_2 bc\}) = \epsilon \Lambda^n \left(\left\| \begin{matrix} 0 & 1 \\ 1 & 0 \end{matrix} \right\| \right).$$

We let the reader to fill in the details necessary to prove that σ extends to a $\mathbb{R}$-J*-isomorphism of j' onto $I^{\mathbb{C}}_{2n,1}$.

The lemma is completely proved.

LEMMA 33. Let a,b and c be three non zero, idempotent, irreducible elements of a $\mathbb{R}$-J*-algebra j such that:

1) a and b are strongly independent;

2) a and c are strongly independent;

3) $\delta_b(j) \cap \delta_c(j) = \{0\}$;

4) $\psi_b(c) - c = \psi_c(b) - b = 0$;

5) $\psi_a\psi_b(c) \neq \{0\}$.

Then $\psi_a\psi_b(\mathcal{J})$ is $\mathbb{R}$-J*-isomorphic to one of the following $\mathbb{R}$-J*-algebras:

$$M(2;\mathbb{H}),\ I_2\ I_2^{\mathbb{C}},\ I_4,\ I_4^{\mathbb{C}}.$$

PROOF. Let us consider the three $\mathbb{R}$-J*-subalgebras of $\mathcal{J}$:

$$I_1 = \psi_a\psi_b(\mathcal{J}) \cap \psi_a\psi_c(\mathcal{J}),$$

$$I_2 = \psi_a\psi_b(\mathcal{J}) \cap \Gamma_c(\mathcal{J}),$$

$$I_3 = \psi_a\psi_c(\mathcal{J}) \cap \Gamma_b(\mathcal{J}).$$

(Each one of them is the intersection of two $\mathbb{R}$-J*-algebras). The assumptions made on a,b and c entail. $\psi_a\psi_b = \psi_b\psi_a$, $\psi_a\psi_c = \psi_c\psi_a$ and $\psi_b\psi_c = \psi_c\psi_b$. The first and second relation are consequences of 3) and of lemma 3; the third one is a trivial consequence of $bb^*cc^* = cc^*bb^*$ and $b^*bc^*c = c^*cb^*b$. The above relations yield $\psi_a\Gamma_c = \Gamma_c\psi_a$, $\psi_b\Gamma_c = \Gamma_c\psi_b$, $\psi_a\Gamma_b = \Gamma_b\psi_a$ and hence $I_1 = \psi_a\psi_b\psi_c(\mathcal{J})$, $I_2 = \psi_a\psi_b\Gamma_c(\mathcal{J})$ and $I_3 = \psi_a\psi_c\Gamma_b(\mathcal{J})$. It follows that $I_i \cap I_j = \{0\}$ whenever $1 \leq i < j \leq 3$. Since $\mathcal{J} = \psi_b(\mathcal{J}) + \Gamma_b(\mathcal{J}) = \psi_c(\mathcal{J}) + \Gamma_c(\mathcal{J})$, we have $\psi_a\psi_b(\mathcal{J}) = I_1 + I_2$ and $\psi_a\psi_c(\mathcal{J}) = I_1 + I_3$. The mapping $\Delta : \psi_a\psi_b(\mathcal{J}) \to \psi_a\psi_b(\mathcal{J})$ defined by $\Delta(x) = 6\{abx\}$ maps bijectively I_1 onto I_2. In fact, since

$$\psi_c(6\{abx\}) = cc^*(ax^*b + bx^*a) + (ax^*b + bx^*a)c^*c -$$

$$- cc^*(ax^*b + bx^*a)c^*c = cc^*bx^*a + ax^*bc^*c =$$

$$= 6\{abx\} - 6\{ab\psi_c(x)\}$$

we have that if $\psi_c(x) = x$ then $\psi_c(6\{abx\}) - 0$; if $\psi_c(x) = 0$ then $\psi_c(6\{abx\}) = 6\{abx\}$. Similarly we can prove that the mapping

$H : \psi_a\psi_c(\mathcal{J}) \to \psi_a\psi_c(\mathcal{J})$, defined by $H(x) = 6\{acx\}$, maps bijectively I_1 onto I_3.

Let x be a non zero, idempotent element of I_1. Then $xa^*x = \delta_b(3\{axx\})$ is an element of $\delta_b(\mathcal{J}) \cap \delta_c(\mathcal{J}) = \{0\}$ If follows that $x^*(ax^*b + bx^*a) = x^*bx^*a = x^*cc^*bc^*cx^*a = 0$. The previous relation and the analogous $(ax^*b + bx^*a)a^* = 0$ give (in view of lemma 13 and $ax^*b + bx^*a \neq 0$) that x and $ax^*b + bx^*a$ are strongly independent. If x is a non zero, idempotent, irreducible element of I_1 and $\tilde{x} = 6\{abx\}$ (an element of I_3), it is trivial to prove that $\psi_c\psi_x = \psi_x\psi_c$ and $\psi_c\psi_{\tilde{x}} = \psi_{\tilde{x}}\psi_c$. Moreover if follows that $\Gamma_c\psi_{\tilde{x}} = \psi_{\tilde{x}}\Gamma_c$ and $\Gamma_c\psi_x = \psi_x\Gamma_c$. Let $\mathcal{B}$ be a subset of I_1 which satisfies 1),...,8) of lemma 29 and let $N(\mathcal{J}) = (\pi_x\psi_x)(\mathcal{J})$ $(x \in \mathcal{B} \cup \tilde{\mathcal{B}} \cup \{a,b\})$. Then

$$N(\mathcal{J}) = (\Gamma_c + \psi_c)N(\mathcal{J}) = N(\mathcal{J}) \cap I_1 + N(\mathcal{J}) \cap I_2.$$

$N(\mathcal{J}) \cap I_1$ and $N(\mathcal{J}) \cap I_2$ are two $\mathbb{R}$-J*-subalgebras of $\psi_a\psi_b(\mathcal{J})$ of the same dimension. Now, if we assume $N(\mathcal{J}) \neq \{0\}$, it follows that $N(\mathcal{J}) \cap I_1 \neq \{0\}$ and hence $N(\mathcal{J})$ has a non zero, idempotent, irreducible (in $\mathcal{J}$) element. If we proceed as in the proof of lemma 29, we obtain a subset $\mathcal{B}$ of $\psi_a\psi_b(\mathcal{J})$ that satisfies 1),...,9) of the same lemma and the following relations.

i) $\mathcal{B} \subset I_1$:

ii) $\psi_a\psi_b(\mathcal{J}) = \Sigma_x\delta_x(\mathcal{J})$ $(x \in \mathcal{B} \cup \tilde{\mathcal{B}})$.

Now let $x, y \in \mathcal{B}$ with $x \neq y$. Since $\delta_b(6\{xya\}) = xa^*y + ya^*x$ and $\delta_c(xa^*y + ya^*x) = xa^*y + ya^*x$ we have $xa^*y + ya^*x \in \delta_b(\mathcal{J}) \cap \delta_c(\mathcal{J}) = 0$ and then (see lemma 29) $xa^*y = \tilde{bx}^*y = x\tilde{y}^*b = 0$. Moreover $\delta_a(6\{xyb\}) = xb^*y + yb^*x = 2xb^*y = 2aa^*xc^*cb^*cc^*ya^*a = 0$. Then we obtain $x\tilde{y}^* = x\tilde{y}^*aa^* + x\tilde{y}^*bb^* = 0$, $\tilde{y}x^* = 0$, $\tilde{x}y^* = a^*a\tilde{x}^*y + b^*b\tilde{x}^*y = 0$, $y^*\tilde{x} = 0$. Assume now that $\mathcal{B}$ has three distinct elements: x, y and z. We have $xy^*z = \tilde{x}\tilde{z}^*y = 0$, $xy^*\tilde{z} = 0$ and hence $xy^* = xy^*zz^* + yx^*\tilde{z}\tilde{z}^* = 0$ The last relations give the contradiction $x = xy^*y + xy^*\tilde{y} = 0$.

The lemma follows now from lemma 31.

Let $\mathcal{J}$ be a J*-algebra. We shall define as <u>height</u> of $\mathcal{J}$, $\mathrm{Hght}(\mathcal{J})$, the least positive integer such that whenever B is a finite subset of

$\mathcal{J} - \{0\}$ with $|B| > \mathrm{Hght}(\mathcal{J})$, then B has two elements h,k such that either $h^*k = k^*h \neq 0$ or $hk^* = kh^* \neq 0$.

It is trivial to prove that if $\mathcal{J}$ is a J*-algebra and $\mathcal{J} \neq \{0\}$, then

$$1 \leqslant \mathrm{Hght}(\mathcal{J}) \leqslant \dim \mathcal{J}.$$

<u>LEMMA 34</u>. Let $\mathcal{J}$ be a $\mathbb{R}$-J*algebra of dimension n and height 1. $\mathcal{J}$ is a $\mathbb{R}$-J*-algebra $\mathbb{R}$-J*-isomorphic to $M(1,n;\mathbb{R})$.

<u>PROOF</u>. By lemma 1 and lemma 5, we have:

i) if $x \in \mathcal{J}$ then $xx^*x \doteq x$;

ii) any non zero, idempotent element of $\mathcal{J}$ is irreducible.

Let $u_1,\ldots,u_n$ be a $\mathbb{R}$-basis of $\mathcal{J}$ whose elements are non zero and idempotent. We define, by an inductive argument:

$$a_1 = u_1$$

$$a_{i+1} = \lambda_{i+1}(u_{i+1} - \frac{1}{2}\Sigma_r(a_r a_r^* u_{i+1} a_r^* a_r + a_r u_{i+1}^* a_r)) \quad (r = 1,\ldots,i)$$

(λ_{i+1} is a coefficient chosen so that a_{i+1} is idempotent).
With an easy induction argument we can prove that the set $\{a_1,\ldots,a_n\}$ is $\mathbb{R}$-linearly independent and satisfies $a_r a_r^* a_s a_r^* a_r + a_r a_s^* a_r = 0$ whenever $1 \leqslant r < s \leqslant n$. Since

$$\psi_{a_i}(a_j) = a_i a_i^* a_j + a_j a_i^* a_i - a_i a_i^* a_j a_i^* a_i = a_j \quad \text{for } i,j = 1,\ldots,n$$

$$a_i a_i^* a_j a_i^* a_i + a_i a_j^* a_i \doteq a_i \quad \text{for } i,j = 1,\ldots,n,$$

we obtain $a_i - 3\{a_i a_i a_j\} \doteq a_i$ whenever $i,j = 1,\ldots,n$ (in particular it is $a_i - 3\{a_i a_i a_j\} = 0$ whenever $1 \leqslant i < j \leqslant n$). Since

$$a_i + a_j \doteq \{(a_i + a_j)(a_i + a_j)(a_i + a_j)\} = a_i + 3\{a_i a_i a_j\} +$$

$$+ 3\{a_j a_j a_i\} + a_j = a_i + 2a_j + 3\{a_j a_j a_i\},$$

we have $a_i \doteq a_i - 3\{a_j a_j a_i\} \doteq a_i + a_j$. Hence $3\{a_r a_r a_s\} \doteq a_s$ whenever $r,s = 1,\dots,n$. Now let $i,r,j = 1,\dots,n$ be three distinct integers. Then

$$a_i + a_j + a_r \doteq \{(a_i + a_j + a_r)(a_i + a_j + a_r)(a_i + a_j + a_r)\} =$$

$$= 3(a_i + a_j + a_r) + 6\{a_i a_j a_r\}$$

and hence $6\{a_i a_j a_r\} \doteq a_i + a_j + a_r$. It follows that there exists a function $\varphi : \mathbb{R}^2 \to \mathbb{R}$ such that

$$\varphi(\lambda,\mu)(a_i + \lambda a_j + \mu a_r) = \{(a_i + \lambda a_j + \mu a_r)(a_i + \lambda a_j + \mu a_r)$$

$$(a_i + \lambda a_j + \mu a_r)\} = (1 + \lambda^2 + \mu^2 + k\lambda\mu)a_i + (\lambda^3 + \lambda + \lambda^2 +$$

$$+ k\lambda\mu)a_j + (\mu^3 + \mu + \mu^2 + k\lambda\mu)a_r$$

and this is possible if, and only if, $k = 0$, that is $\{a_i a_j a_r\} = 0$. For $i = 1,\dots,n$; let us denote $E_i = \|a_{ih}\| \in \mathcal{M}(1,n;\mathbb{R})$, the matrix whose entries are zero whenever $h = 1,\dots,n$ and $h \neq i$, and $a_{ii} = 1$. The mapping $\sigma : \{a_1,\dots,a_n\} \to \mathcal{M}(1,n.\mathbb{R})$ defined by $\sigma(a_i) = E_i$, $i = 1,\dots,n$ extends to a $\mathbb{R}$-J*-isomorphism of $\mathcal{J}$ onto $\mathcal{M}(1,n;\mathbb{R})$.

<u>LEMMA 35</u>. Let $\mathcal{J}$ be a J*-algebra, of height h, and let $\{x_1,\dots,x_h\}$ be a subset whose elements are non zero, idempotent, irreducible pairwise strongly independent. The subsets:

$$I_i = \{x : x \in \mathcal{J} \text{ such that } \psi_{x_i}(x) = x, \psi_{x_1}(x) = \dots = \psi_{x_{i-1}}(x) =$$

$$= \psi_{x_{i+1}}(x) = \dots = \psi_{x_h}(x) = \delta_{x_i}(x) = 0\}$$

are $\mathbb{R}$-J*-isomorphic to J*-subalgebras of $\mathcal{J}$ whose height is 1. If $I_i \neq \{0\}$ then any non zero, idempotent element of I_i which is irreducible in I_i is irreducible in $\mathcal{J}$, $i = 1,\dots,h$.

<u>PROOF</u>. Whenever $x \in I_i$, $i = 1,\dots,h$, we have the following relations:

i) $\quad x_i x^* x_i = 0$

ii) $\quad x = x_i x_i^* x + x x_i^* x_i$,

iii) $\quad x x_j^* = x_j x^* = x_j^* x = x^* x_j = 0$ whenever $j = 1,\dots,h$ and $j \neq i$.

In fact, since $\delta_{x_i}(x) = x_i x_i^* x x_i^* x_i$, we have i). ii) follows from i) and $x = \psi_{x_i}(x) = x_i x_i^* x + x x_i^* x_i - x_i x_i^* x x_i^* x_i$. Since $0 = \psi_{x_j}(x)$, applying δ_{x_j}, we obtain $x_j x_j^* x x_j^* x_j = 0$ and hence $x_j^* x x_j^* = 0$. iii) follows from $x_j^* x = x^* x_j^* = 0$ (obtained from $0 = x_j x_j^* x + x x_j^* x_j$). Using i), ii) and iii), we have, for $x \in I_i$, $j = 1,\dots,h$ with $j \neq i$,

$$\psi_{x_i}(xx^*x) = xx^*x, \quad \delta_{x_i}(xx^*x) = 0, \quad \psi_{x_j}(xx^*x) = 0.$$

It follows that I_i is a J*-algebra.

Now let $i,j = 1,\dots,h$, with $i \neq j$. Since $\mathcal{J}$ is irreducible $\psi_{x_i}\psi_{x_j}(\mathcal{J}) \neq 0$ and hence there exists a non zero, idempotent element $c \in \psi_{x_i}\psi_{x_j}(\mathcal{J})$ such that $6\{x_i x_j c\} = \epsilon c$ with $\epsilon \in \{1, -1\}$. Whenever $x \in I_i$ we have $6\{x x_i c\} = x x_i^* c + c x_i^* x \in I_j$. In fact we have:

$$\psi_{x_j}(x x_i^* c + c x_i^* x) = x_j x_j^* c x_i^* x + x x_i^* c x_j^* x_j = x x_i^* c + c x_i^* x \quad \text{(see iii))};$$

$$\delta_{x_j}(x x_i^* c + c x_i^* x) = 0 \quad \text{(see iii))};$$

$$\psi_{x_i}(x x_i^* c + c x_i^* x) = 0 \quad \text{(see ii))};$$

$$\psi_{x_r}(x x_i^* c + c x_i^* x) = 0 \text{ whenever } r = 1,\dots,h \text{ with } r \neq i,j \quad \text{(see ii) and iii))}.$$

Let $\vartheta : I_j \to I_i$ and $\varphi : I_i \to I_j$ be the two mappings defined by $\vartheta(x) = 6\{x_j x c\}$ and $(x) = 6\{x_i x c\}$, respectively. We have the relations:

$\varphi\vartheta(x) = \epsilon(xx_j^*c + cx_j^*x)x_1^*c + \epsilon cx_1^*(xx_j^*c + cx_j^*x) = \epsilon xx_j^*cx_1^*c +$

$+ \epsilon cx_1^*cx_j^*x = xx_j^*x_j + x_jx_j^*x = x$ whenever $x \in I_j$;

$\vartheta\varphi(x) = x$ whenever $x \in I_1$;

$\varphi(x)\varphi(x)^*\varphi(x) = xx_1^*cc^*x_1x^*xx_1^*c + cx_1^*xx^*x_1c^*cx_1^*x = xx_1^*x_1x^*xx_1^*c +$

$+ cx_1^*xx^*x_1x_1^*c = \varphi(xx^*x)$.

Hence I_1 is J*-isomorphic to I_j.
iii) implies that any non zero, idempotent, irreducible element of I_1 is strongly independent of $x_2,\dots,x_h$. Hence, since $\mathrm{Hght}(\mathcal{J}) = h$, I_1 cannot have two non zero, idempotent, strongly independent elements. It follows that $\mathrm{Hght}(I_1) = \dots, \mathrm{Hght}(I_h) = 1$. Now we have only to prove that any irreducible element $x \in I_1$ is irreducible in $\mathcal{J}$ too. Since

$$\mathcal{J} = \delta_{x_1}(\mathcal{J}) + \Sigma_j \psi_{x_j}(\mathcal{J}) + I_1 \quad (j = 2,\dots,h)$$

let us consider two cases:

I) Assume $y \in \psi_{x_1}(\mathcal{J})$. We have $\Gamma_{x_1}(3\{xxy\}) = xy^*x$. Since $x_1x_1^*xy^*x = xy^*x - xx_1^*x_1y^*x = 0$, we obtain $x_1^*(xy^*x) = 0$. In analogy, we can prove that $(xy^*x)x_1^* = 0$. iii) implies that $x_j^*xy^*x = xy^*xx_j^* = 0$ whenever $j = 2,\dots,h$. It follows that xy^*x is strongly independent from $x_1,\dots,x_h$ and hence is zero.

II) Assume $y \in \psi_{x_j}(\mathcal{J})$, with $j \neq 1$, we have $xy^*x = x(x_j^*x_jy^* + y^*x_jx_j^* - x_j^*x_jy^*x_jx_j^*)x = 0$.

The last part of the lemma now follows trivially by linearity.

LEMMA 36. Let $\mathcal{J}$ be an irreducible $\mathbb{R}$-J*-algebra of height $h \geq 2$, and let $\{x_1,\dots,x_h\}$ be a subset of $\mathcal{J}$ whose elements are non zero, idempotent, irreducible and pairwise strongly independent. If $\psi_{x_1}\psi_{x_2}(\mathcal{J})$

does not contain any non zero, idempotent element which is irreducible in j, then j is $\mathbb{R}$-J*-isomorphic to one of the following $\mathbb{R}$-J*-algebras.

$$I_{2,m}\ ,\quad S(n;\mathbb{R})\ ,\quad S(n;\mathbb{C})\ ,\quad H(n;\mathbb{C})\ ,\quad H(n;\mathbb{H})\ ,\quad H^*(n;\mathbb{H})$$

$(n \geqslant 2$ and $m \geqslant 1)$.

PROOF. First let us prove that

$$j = \Sigma_i \delta_{x_i}(j) + \Sigma_{ij} \psi_{x_i}\psi_{x_j}(j) \qquad (i,j = 1,\ldots,h \text{ with } i \neq j).$$

Assume the above decomposition is false. Then there exists a non zero, idempotent, irreducible element $x \in I_2$ (I_2 is defined in the proof of lemma 35). x_1, x_2 and x then satisfy the assumption of lemma 33 and this implies the existence, in $\psi_{x_1}\psi_{x_2}(j)$ of a non zero, idempotent element which is irreducible in j, a contradiction. If $h = 2$, our lemma is a consequence of lemma 28 and corollary 6. Assume now $h \geqslant 3$ and let us prove that $\dim \psi_{x_1}\psi_{x_2}(j) = 1, 2$ or 4.

By lemma 28, it suffices to prove our assertion assuming $\dim \delta_{x_1}(j) = \dim \delta_{x_2}(j) = 1$ and $\dim \psi_{x_1}\psi_{x_2}(j) \geqslant 3$. Let c_1, c_2, c_3 be three non zero, idempotent, irreducible elements of $\psi_{x_1}\psi_{x_2}(j)$ satisfying 5) of lemma 11 and $6\{x_1 x_2 c_i\} = \epsilon c_i$, where $i = 1,2,3$ and $\epsilon \in \{1, -1\}$. Now let $k \in \psi_{x_3}\psi_{x_2}(j)$ be a non zero, idempotent irreducible (in $\psi_{x_3}\psi_{x_2}(j)$) element such that $6\{x_2 x_3 k\} = \mu k$ with $\mu \in \{1, -1\}$ (see lemma 9). Moreover consider the elements $t = 6\{c_1 x_2 k\} = c_1 x_2^* k + k x_2^* c_1 \in \psi_{x_1}\psi_{x_3}(j)$, and $q = 6^3\{x_3\{c_2 x_1 t\}\{c_3 x_2 k\}\} = x_1 c_1^* c_2 c_3^* x_2 + x_2 c_3^* c_2 c_1^* x_1 \in \psi_{x_1}\psi_{x_2}(j)$. Since $c_1 c_2^* c_3 \neq 0$ we have $q \neq 0$. Now it is easy to prove that $c_i^* q + q^* c_i = c_i q^* + q c_i^* = 0$ whenever $i = 1,2,3$ and hence that $\dim \psi_{x_1}\psi_{x_2}(j) \geqslant 4$ (see lemma 9). Using an argument similar to the one used in the proof of lemma 30, we can prove that $\dim \psi_{x_1}\psi_{x_2}(j) > 4$ is impossible. Hence we have only to consider the following five cases:

I) $\dim\delta_{x_1}(j) = 1$ and $\dim\psi_{x_1}\psi_{x_2}(j) = 1$;

II) $\dim\delta_{x_1}(j) = 1$ and $\dim\psi_{x_1}\psi_{x_2}(j) = 2$,

III) $\dim\delta_{x_1}(j) = 1$ and $\dim\psi_{x_1}\psi_{x_2}(j) = 4$;

IV) $\dim\delta_{x_1}(j) = 2$ and $\dim\psi_{x_1}\psi_{x_2}(j) = 2$;

V) $\dim\delta_{x_1}(j) = 3$ and $\dim\psi_{x_1}\psi_{x_2}(j) = 4$.

Now let us prove that there exist a subset of j, $\{c_{ij} = c_{ji} : 1 \leqslant i < j \leqslant h\}$, and h coefficients $\epsilon_1, \ldots, \epsilon_h \in \{1, -1\}$ such that

i) c_{ij} is a non zero, idempotent element of $\psi_{x_i}\psi_{x_j}(j)$, whenever $1 \leqslant i < j \leqslant h$;

ii) $6\{c_{ij}x_jc_{jr}\} = \epsilon_j c_{ir}$ whenever $i,j,r = 1,\ldots,h$ with i,j,r distinct integers;

iii) $6\{x_ic_{ij}x_j\} = \epsilon_i\epsilon_j c_{ij}$ whenever $i,j = 1,\ldots,h$ with $i \neq j$.

Let us define the above elements using a finite induction argument. By lemma 9, whenever $i = 1,\ldots,h-1$, we can choose a non zero, idempotent element $c_{i\,i+1} = c_{i+1\,i} \in \psi_{x_i}\psi_{x_{i+1}}(j)$ such that $6\{x_ic_{i\,i+1}x_{i+1}\} = \mu_i c_{i\,i+1}$ with $\mu_i \in \{1, -1\}$ Let us define

$$\epsilon_1 = 1 \text{ and } \epsilon_{i+1} = \mu_i\epsilon_i;$$

$$c_{ij} = c_{ji} = \epsilon_{i+1} \ldots \epsilon_{j-1}(c_{i\,i+1}x^*_{i+1}c_{i+1\,i+2}x^*_{i+2} \ldots x^*_{j-1}c_{j-1\,j} + c_{j\,j-1}x^*_{j-1} \ldots x^*_{i+1}c_{i+1\,i}) \text{ whenever } 1 \leqslant i < j \leqslant h \text{ with }$$

$$j - i \geqslant 2.$$

It is trivial to prove that, whenever $1 \leqslant i < j \leqslant h$, we have

$\epsilon_j c_{i\,j+1} = 6\{c_{ij} x_j c_{j\,j+1}\}$ and hence, using a simple induction argument, $c_{ij} \in \mathit{j}$ whenever $1 \leqslant i < j \leqslant h$.

i) follows from

$$\psi_{x_i}(c_{ij}) = x_i x_i^* c_{ij} + c_{ji} x_i^* x_i - x_i x_i^* c_{ij} x_i^* x_i = \epsilon_{i+1} \cdots \epsilon_{j-1}$$

$$(x_i x_i^* c_{ij} x_{i+1}^* \cdots x_{j-1}^* c_{j-1\,j} + c_{j\,j-1} x_{j-1}^* \cdots x_{i+1}^* c_{i+1\,i} x_i^* x_i) = c_{ij}$$

(let $\epsilon_{i+1} \cdots \epsilon_{j-1}$ denote the integer 1 whenever $j-1 = 1$) and the analogous relation $\psi_{x_j}(c_{ij}) = c_{ij}$.

To prove ii) we consider two cases:

1) $i < j < r$. We have $6\{c_{ij} x_j c_{jr}\} = c_{ij} x_j^* c_{ir} + c_{rj} x_j^* c_{ji} =$

$$\epsilon_{i+1} \cdots \epsilon_{j-1} \epsilon_{j+1} \cdots \epsilon_{r-1} (c_{i\,i+1} x_{i+1}^* \cdots c_{j-1\,j} x_j^*$$

$$c_{j\,j+1} x_{j+1}^* \cdots x_{r-1}^* c_{r-1\,r} + c_{r\,r-1} x_{r-1}^* \cdots c_{i+1\,i}) = \epsilon_j c_{ir}.$$

2) $i < r < j$. We have (see lemma 9): $6\{c_{ij} x_j c_{jr}\} =$

$$\epsilon_{i+1} \cdots \epsilon_{j-1} \epsilon_{r+1} \cdots \epsilon_{j-1} (c_{i\,i+1} \cdots c_{j-1\,j} x_j^* c_{j\,j-1} x_{j-1}^* \cdots x_{r+1}^*$$

$$c_{r+1\,r} + c_{r\,r+1} x_{r+1}^* \cdots c_{j-1\,j} x_j^* \cdots x_{i+1}^* c_{i+1\,i}) = \epsilon_{i+1} \cdots \epsilon_{r-1}$$

$$\epsilon_j (c_{i\,i+1} x_{i+1}^* \cdots x_{r-1}^* c_{r-1\,r} + c_{r\,r-1} \cdots c_{i+1\,i}) = \epsilon_j c_{ir}.$$

The other possible cases are always reducible to case 1) or 2). iii) is trivially proved using the definition of $\epsilon_1, \ldots, \epsilon_h$. Set $c_{ii} = \epsilon_i x_i$, $i = 1, \ldots, h$. It follows that, for $i, j = 1, \ldots, h$, with $i \neq j$ and $x \in \psi_{x_1} \psi_{x_2}(\mathit{j})$, $6\{c_{i1} c_{2j} x\} = c_{i1} x^* c_{2j} + c_{j2} x^* c_{1i} \in \psi_{x_i} \psi_{x_j}(\mathit{j})$ Hence we can define a mapping $\Delta_{ij} : \psi_{x_1} \psi_{x_2}(\mathit{j}) \to \psi_{x_i} \psi_{x_j}(\mathit{j})$ by $\Delta_{ij}(x) = c_{i1} x^* c_{2j} + c_{j2} x^* c_{1i}$. Since $6\{c_{i1} c_{2j} \Delta_{ij}(x)\} = c_{1i} c_{i1}^* x c_{2j}^* c_{j2} +$

$c_{2j}c^*_{j2}xc^*_{1i}c_{i1} = c_{11}c^*_{11}xc^*_{22}c_{22} + c_{22}c^*_{22}xc^*_{11}c_{11} = x$, the $\mathbb{R}$-linear mapping Δ_{ij} is a $\mathbb{R}$-isomorphism, in fact, a $\mathbb{R}$-J*-isomorphism. Let $\{a_1 = c_{11},\ldots,a_p\}$ be a $\mathbb{R}$-basis of $\delta_{x_1}(j)$ given by lemma 11. Whenever $x \in \delta_{x_1}(j)$, with $x = \lambda_1 a_1 + \ldots + \lambda_p a_p$, the element

$$c_{i1}x^*c_{1i} = \begin{cases} \lambda_1 a_1 - \lambda_2 a_2 - \ldots - \lambda_p a_p & \text{if } i = 1 \\ \delta_{x_1}(3\{c_{i1}c_{i1}x\}) & \text{if } i = 2,\ldots,h \end{cases}$$

belongs to $\delta_{x_i}(j)$. Hence the $\mathbb{R}$-linear mapping $\Delta_{1i} : \delta_{x_1}(j) \to \delta_{x_i}(j)$, defined by $\Delta_{1i}(x) = c_{i1}x^*c_{1i}$, is an isomorphism, more a J*-isomorphism (since $c_{1i}(\Delta_{1i}(x))^*c_{i1} = x$).

Let $\{c_1 = c_{12}, c_2,\ldots,c_q\}$ be a basis of $\psi_{x_1}\psi_{x_2}(j)$ given by lemma 11.

Whenever $i, j = 1,\ldots,h$ with $i \neq 0$, we have

$$\Delta_{1j}\Delta_{ji}(c_r) = c_{1i}c^*_{i2}c_r c^*_{1j}c_{j2} + c_{2j}c^*_{j1}c_r c^*_{2i}c_{i1} = c_{12}c^*_{22}c_r c^*_{12}c_{22} +$$

$$+ c_{22}c^*_{21}c_r c^*_{22}c_{21} = E(c_r) = \pm c_r.$$

It follows that

$$j = \Sigma_i \Delta_{1i}\delta_{x_1}(j) + \Sigma_{rs}\Delta_{rs}\psi_{x_1}\psi_{x_2}(j) \quad (1 \leq i \leq h;\ 1 \leq r < s \leq h)$$

and hence

$$(\cup_i \Delta_{1i}(\{a_1,\ldots,a_p\})) \cup (\cup_{rs}\Delta_{rs}(\{c_1,\ldots,c_q\})) \quad (1 \leq i \leq h;\ 1 \leq r < s \leq h)$$

is a $\mathbb{R}$-basis of j.

Whenever $i, j = 1,\ldots h$, let us denote by M_{ij} the square matrix of order h $\|a_{rs}\|$ whose entries are

$$a_{rs} = \begin{cases} 1 & \text{whenever } r = i \text{ and } s = j \\ 0 & \text{elsewere.} \end{cases}$$

The set $\{M_{11},\ldots,M_{hh}\}$ is a subset of $\mathcal{S}(h;\mathbb{R})$, $\mathcal{S}(h;\mathbb{C})$, $\mathcal{H}(h;\mathbb{C})$, $\mathcal{H}(h;\mathbb{H})$, whose elements are non zero, idempotent, irreducible and two by two strongly independent. ($\{iM_{11},\ldots,iM_{hh}\}$ is such a set for $\mathcal{H}^*(h;\mathbb{H})$). If H denotes any of the above $\mathbb{R}$-J*-algebras we have

$$H = \Sigma_i \delta_{M_{ii}}(H) + \Sigma_{rs} \psi_{M_{rr}} \psi_{M_{ss}}(H) \quad (1 \leqslant i \leqslant h;\ 1 \leqslant r < s \leqslant h)$$

$$(H = \Sigma_p \delta_{iM_{pp}}(H) + \Sigma_{rs} \psi_{iM_{rr}} \psi_{iM_{ss}}(H) \quad (1 \leqslant p \leqslant h;\ 1 \leqslant r < s \leqslant h)$$

where $H = \mathcal{H}^*(h;\mathbb{H})$). The set $\{M_{ij} + M_{ji} : 1 \leqslant i < j \leqslant h\}$ is a subset of $\mathcal{S}(h;\mathbb{R})$, $\mathcal{S}(h;\mathbb{C})$, $\mathcal{H}(h;\mathbb{C})$, $\mathcal{H}(h;\mathbb{H})$ which satisfies the properties i), ii) and iii) when $\epsilon_1 = \ldots = \epsilon_h = 1$. ($\{iM_{rs} + iM_{sr} : 1 \leqslant r < s \leqslant h\}$ is a subset of $\mathcal{H}^*(h;\mathbb{H})$ satisfying the same properties when $\epsilon_1 = \ldots = \epsilon_h = 1$). Let us denote Δ_{ji} and Δ_{ij} the $\mathbb{R}$-J*-algebras isomorphisms on H defined in a similar way to the ones in $\mathcal{J}$ and denoted by the same symbol. To prove that there exists a $\mathbb{R}$-J*-isomorphism of $\mathcal{J}$ onto one of the J*-algebras $\mathcal{S}(h;\mathbb{R}),\ldots,\mathcal{H}^*(h;\mathbb{H})$ we prove a stronger proposition i.e. we prove the existence of a J*-isomorphism σ such that:

1) $\sigma\Delta_{1i}(x) = \Delta_{1i}\sigma(x)$ whenever $x \in \delta_{x_1}(\mathcal{J})$;

2) $\sigma\Delta_{ij}(x) = \Delta_{ij}\sigma(x)$ whenever $x \in \psi_{x_1}\psi_{x_2}(\mathcal{J})$, $i, j = 1,\ldots,h$ with $i \neq j$.

Let x,y,z be three arbitrary elements of $\psi_{x_1}\psi_{x_2}(\mathcal{J})$ and α,β,γ be three arbitrary elements of $\delta_{x_1}(\mathcal{J})$. Moreover let $E = \Delta_{12}\Delta_{21} : \psi_{x_1}\psi_{x_2}(\mathcal{J}) \to \psi_{x_1}\psi_{x_2}(\mathcal{J})$. Now we obtain a convenient formula to express $6\{abc\}$, whenever a,b,c are chosen in the images of Δ_{ij}, Δ_{rs} and Δ_{pq} with $i,j,r,s,p,q = 1,\ldots,h$.

Clearly $6\{abc\} = 0$ unless we are in one of the following case: (hereafter $i,j,r,s = 1,\dots,h$ are four distinct integers)

1) $6\{\Delta_{ij}(x)\Delta_{jr}(y)\Delta_{rs}(z)\} = \Delta_{is}(xE(y)^*zc^*_{22}c_{22} + c_{22}c^*_{22}zE(y)^*x);$

2) $6\{\Delta_{ii}(\alpha)\Delta_{ir}(x)\Delta_{rs}(y)\} = \Delta_{is}(aE(x)^*yc^*_{22}c_{22} + c_{22}c^*_{22}yE(x)^*a)$ (where

$$a = 6\{\alpha c_{22}c_{12}\});$$

3) $6\{\Delta_{ij}(x)\Delta_{ji}(y)\Delta_{is}(z)\} = \Delta_{is}(xE(y)^*zc^*_{22}c_{22} + c_{22}c^*_{22}zE(y)^*x +$

$$+ E(y)x^*zc^*_{22}c_{22} + c_{22}c^*_{22}zx^*E(y));$$

4) $6\{\Delta_{ij}(x)\Delta_{jr}(y)\Delta_{ri}(z)\} = \Delta_{ii}(6\{xE(y)^*zc^*_{22}c_{22} + c_{22}c^*_{22}zE(y)^*x)c^*_{22}c_{21}\})$

$+ \Delta_{jj}(6\{(xE(z)^*yc^*_{22}c_{22} + c_{22}c^*_{22}yE(z)^*x)c^*_{22}c_{21}\}) +$

$+ \Delta_{rr}(6\{(yE(x)^*zc^*_{22}c_{22} + c_{22}c^*_{22}zE(x)^*y)c^*_{22}c_{21}\});$

5) $6\{\Delta_{ij}(x)\Delta_{jj}(\alpha)\Delta_{jr}(y)\} = \Delta_{is}(xE(a)^*zc^*_{22}c_{22} + c_{22}c^*_{22}zE(a)^*x);$

6) $6\{\Delta_{ii}(\alpha)\Delta_{ii}(\beta)\Delta_{is}(x)\} = \Delta_{is}(6\{\alpha\beta x\});$

7) $6\{\Delta_{ii}(\alpha)\Delta_{ir}(x)\Delta_{ri}(y)\} = \Delta_{ii}(6\{(aE(x)^*yc^*_{22}c_{22} + c_{22}c^*_{22}yE(x)^*a)c^*_{22}c_{21}$

$+ 6\{(ay^*E(x)c^*_{22}c_{22} + c_{22}c^*_{22}E(x)y^*a)c^*_{22}c_{21}\}) +$

$+ \Delta_{rr}(6\{(yE(x)^*ac^*_{22}c_{22} + c_{22}c^*_{22}aE(x)^*y)c^*_{22}c_{21}\});$

8) $6\{\Delta_{ii}(\alpha)\Delta_{ir}(x)\Delta_{rr}(\beta)\} = \Delta_{ir}(aE(x)^*bc^*_{22}c_{22} + c_{22}c^*_{22}bE(x)^*a)$

(where $b = 6\{\beta c_{22}c_{21}\}$);

9) $6\{\Delta_{ij}(x)\Delta_{ji}(y)\Delta_{ij}(z)\} = \Delta_{ij}(xE(y)^*zc^*_{22}c_{22} + c_{22}c^*_{22}zE(y)^*x +$

$+ E(x)y^*E(z)c^*_{22}c_{22} + c_{22}c^*_{22}E(z)y^*E(x) + E(y)x^*zc^*_{22}c_{22} +$

$+ c_{22}c^*_{22}zx^*E(y) + yE(x)^*E(z)c^*_{22}c_{22} + c_{22}c^*_{22}E(z)E(x)^*y +$

$+ xz^*E(y)c_{22}^*c_{22} + c_{22}c_{22}^*E(y)z^*x + E(x)E(z)^*yc_{22}^*c_{22} +$

$+ c_{22}c_{22}^*yE(z)^*E(x));$

10) $6\{\Delta_{ij}(x)\Delta_{jj}(\alpha)\Delta_{ji}(y)\} = \Delta_{ii}(xE(a)^*yc_{22}^*c_{22} + c_{22}c_{22}^*yE(a)^*x) +$

$+ \Delta_{jj}(6\{(zE(x)^*ac_{22}^*c_{22} + c_{22}c_{22}^*aE(x)^*z)c_{22}^*c_{21} + aE(z)^*xc_{22}^*c_{22} +$

$+ c_{22}c_{22}^*xE(z)^*a);$

11) $6\{\Delta_{ii}(\alpha)\Delta_{ii}(\beta)\Delta_{ii}(\gamma)\} = \Delta_{ii}(6\{\alpha\beta\gamma\}).$

It is evident that σ satisfies i) and ii) if, and only if, there exists a $\mathbb{R}$-J*-isomorphism

$$\sigma : \delta_{x_1}(j) + \delta_{x_2}(j) + \psi_{x_1}\psi_{x_2}(j) \to \delta_{M_{11}}(H) + \delta_{M_{22}}(H) + \psi_{M_{11}}\psi_{M_{22}}(H)$$

(onto $\delta_{iM_{11}}(H) + \delta_{iM_{22}}(H) + \psi_{iM_{11}}\psi_{iM_{22}}(H)$, when $H = H^*(h;\mathbb{H})$), such that:

a) $\sigma(c_{11}) = M_{11}$ $(= iM_{11});$

b) $\sigma(c_{22}) = M_{22}$ $(= iM_{22});$

c) $\sigma(c_{12}) = M_{12} + M_{21}$ $(= iM_{12} + iM_{21});$

d) $\sigma E = E\sigma;$

e) $\sigma(xy^*zc_{22}^*c_{22} + c_{22}c_{22}^*zy^*x) = \sigma(x)\sigma(y)^*\sigma(z)\sigma(c_{22})^*\sigma(c_{22}) +$

$+ \sigma(c_{22})\sigma(c_{22})^*\sigma(z)\sigma(y)^*\sigma(x)$ whenever $x,y,z \in \psi_{x_1}\psi_{x_2}(j)$.

Now if we identify, naturally, $\delta_{M_{11}}(H) + \delta_{M_{22}}(H) + \psi_{M_{11}}\psi_{M_{22}}(H)$ with $S(2,\mathbb{R})$, $S(2,\mathbb{C})$, $H(2,\mathbb{C})$, $H(2,\mathbb{H})$, depending on H ($\delta_{iM_{11}}(H) + \delta_{iM_{22}}(H) + \psi_{iM_{11}}\psi_{iM_{22}}(H)$ with $H^*(2;\mathbb{H})$), the $\mathbb{R}$-J*-isomorphism defined in lemma 29 satisfies a),...,e).

The proof is now complete.

LEMMA 37. Let $\mathcal{J}$ be a $\mathbb{R}$-J*-algebra irreducible of height $h \geq 2$, and let $\{x_1,\dots,x_h\}$ be a subset of $\mathcal{J}$ whose elements are non zero, idempotent, irreducible and pairwise strongly independent. Moreover let $\mathcal{B}$ be a subset of $\psi_{x_1}\psi_{x_2}(\mathcal{J})$ given by lemma 29 and assume that:

1) $\mathcal{J} \neq \delta_{x_1}(\mathcal{J}) + \delta_{x_2}(\mathcal{J}) + \psi_{x_1}\psi_{x_2}(\mathcal{J})$;

2) $|\mathcal{B}| = 1$.

Then $\mathcal{J}$ is $\mathbb{R}$-J*-isomorphic to one of the following J*-algebras:

$$M(h,p;\mathbb{R}), \quad M(h,p;\mathbb{C}), \quad M(h,p;\mathbb{H})$$

whenever $2 \leq h \leq p$ with $(p,h) \neq (2,2)$.

PROOF. Let x be the unique element of $\mathcal{B}$ and set $\tilde{x} = 6\{x_1 x_2 x\}$. Moreover whenever $3 \leq i \leq h$, let $c_i \in \psi_{x_2}\psi_{x_i}(\mathcal{J})$ be a non zero, idempotent element such that $6\{x_2 x_i c_i\} = \epsilon_i c_i$ with $\epsilon_i \in \{1, -1\}$. Now let us denote $x_2 = x$, $\tilde{x}_2 = \tilde{x}$ and $x_i = 6\{x x_2 c_i\}$ and $\tilde{x}_i = 6\{\tilde{x} x_2 c_i\}$ whenever $3 \leq i \leq h$. It is trivial to verify that $x_2,\dots,x_h$, $\tilde{x}_2,\dots,\tilde{x}_h$ are non zero, idempotent, irreducible elements of $\mathcal{J}$ and that $\psi_{x_1}\psi_{x_i}(\mathcal{J}) = \delta_{x_i}(\mathcal{J}) + \delta_{\tilde{x}_i}(\mathcal{J})$ (see lemma 33). Moreover x_i and $\tilde{x}_j$ are strongly independent whenever $2 \leq i, j \leq h$. Let

$$I = \{ y : y \in \mathcal{J} \text{ such that } \psi_{x_1}(y) = y, \delta_{x_1}(y) = 0 \text{ and } \psi_{x_i}(y) = 0$$
$$\text{whenever } i = 2,\dots,h\}.$$

If $I \neq \{0\}$, I is a J*-algebra of height 1 (see lemma 35) and its non zero, idempotent, irreducible elements are irreducible in $\mathcal{J}$. Assume $I \neq \{0\}$ and let y be one of its non zero, idempotent elements y is strongly independent from x_2 and hence we have $0 = x_2^* y = x^* x_1 \tilde{x}^* y + \tilde{x}^* x_1 \tilde{x}^* y$ and $0 = x x^* x_1 \tilde{x}^* y = x_1 \hat{x}^* y = x_1^* x_1 \tilde{x}^* y$. Consequently we have $\tilde{x}^* y = \tilde{x}^* x_1 x_1^* y$ and $\widetilde{xx}^* y = \widetilde{xx}^* x_1 x_1^* y$. We have also $x x^* y = x x^* x_1 x_1^* y$, thus $\widetilde{xx}^* y + x x^* y = (\widetilde{xx}^* x_1 + x x^* x_1) x_1^* y = x_1 x_1^* y$. In analogy we prove that $y\tilde{x}^*\tilde{x} + y x^* x = y x_1^* x_1$ and then, since $\delta_{x_1}(y) = \delta_x(y) =$

$\delta_{\tilde{x}}(y) = 0$, $y = xx^*y + yx^*x + \widetilde{xx}^*y + y\tilde{x}^*\tilde{x} = \psi_x(y) + \psi_{\tilde{x}}(y)$. The last relation entails $\tilde{x}^*yx^* = x^*y\tilde{x}^* = x^*yx^* = \tilde{x}^*y\tilde{x}^* = 0$. We have $y = \psi_x(y) + \psi_{\tilde{x}}(y) = 2y - (2y - \psi_x(y) + \psi_{\tilde{x}}(y)) = 2y - \Gamma_x(y) - \Gamma_{\tilde{x}}(y)$, and hence $y = \Gamma_x(y) + \Gamma_{\tilde{x}}(y)$. It follows that

$$I = (I \cap \Gamma_x(j)) \cup (I \cap \Gamma_{\tilde{x}}(j))$$

(where $I \cap \Gamma_x(j)$ and $I \cap \Gamma_{\tilde{x}}(j)$ are two J*-algebras). We shall prove that one of the above J*-algebras is zero. Let $y \in I \cap \Gamma_{\tilde{x}}(j)$ and $z \in I \cap \Gamma_x(j)$ be two idempotent elements. We have. $6\{\tilde{x}x_2z\} = x_2\tilde{x}^*z + z\tilde{x}^*x_2$ and then $6\{yx_2(\tilde{x}_2x^*z + z\tilde{x}^*x_2)\} = x_2x_2^*\tilde{x}z^*y + yz^*\tilde{x}x_2^*x_2 = \tilde{x}z^*y + yz^*\tilde{x}$. Since $\delta_{x_1}(\tilde{x}z^*y + yz^*\tilde{x}) = \delta_{x_2}(\tilde{x}z^*y + yz^*\tilde{x}) = \tilde{x}z^*y + yz^*\tilde{x}$ we have $\tilde{x}z^*y + yz^*\tilde{x} = 0$ and then $yz^* = z^*y = 0$. Since I is a J*-algebra of height 1, the last condition is possible if and only if $y = 0$ or $z = 0$. Let us assume $I \cap \Gamma_x(j) = \{0\}$ and let $\{x_{h+1},\dots,x_{h+p}\}$ be a subset of $I = I \cap \Gamma_{\tilde{x}}(j)$ whose elements are non zero, idempotent, irreducible and such that:

i) $\delta_{x_i}(x_j) = 0$ whenever $h < i, j \leq p$ with $i \neq j$;

ii) $\{x_{h+1},\dots,x_{h+p}\}$ is a minimal set.

It is trivial to prove that $I = \Sigma_i \delta_{x_i}(j)$ $(i = h+1,\dots,h+p)$.

Since $\tilde{x}_i^*x_j = (\tilde{x}x_2^*c\, x_i^*x_i + x_ix_i^*c\, x_i^*\tilde{x}_2)x_j^* = 0$ and $x_jx_i^* = 0$ whenever $2 \leq i \leq h$ and $h+1 \leq j \leq p$, it follows that x and $\tilde{x}$ are strongly independent if $2 \leq i \leq h$ and $2 \leq j \leq p$. Let us define the linear mapping $\varphi_{ij} : \delta_{x_1}(j) \to j$, $i = 2,\dots,h$ and $j = 2,\dots,p$, by $\varphi_{ij}(x) = x_ix^*x_j + x_jx^*x_i$. We have

$$j = \Sigma_{ij}\, \varphi_{ij}(\delta_{x_1}(j)) \quad (i = 1,\dots,h \text{ and } j = 1,\dots,p).$$

Whenever we choose $i,j,r = 1,2,\dots,h$ and $t,u,v = 1,\dots,p$ and $x,y,z \in \delta_{x_1}(j)$, we have.

(1) $6\{\varphi_{it}(x)\varphi_{ju}(y)\varphi_{rv}(z)\} = \delta_{tu}\delta_{jr}\varphi_{iv}(xy^*z + zy^*x) +$

$+ \delta_{tv}\delta_{rj}\varphi_{iu}(xz^*y + yz^*x) + \delta_{ut}\delta_{ir}\varphi_{jv}(yx^*z + zx^*y)$

($\delta_{mn} = 1$ if $m = n$, 0 otherwise). Using the notation established in the proof of lemma 36, if $\varphi_{ij} : \delta_{M_{11}}(M(h,p;\mathbb{K})) \to M(h,p;\mathbb{K})$ is the mapping defined by $\varphi_{ij}(x) = M_{i1}x^*M_{1j}$ ($\mathbb{K} = \mathbb{R}$, $\mathbb{C}$ or $\mathbb{H}$), it is trivial to prove that $M(h,p;\mathbb{K}) = \Sigma_{ij}\varphi_{ij}(\delta_{M_{11}}(M(h,p;\mathbb{K})))$ ($i = 1,\dots,h$ and $j = 1,\dots,p$) and φ_{ij} satisfy (1).

Lemma 34 entails that δ_{x_1} is $\mathbb{R}$-J*-isomorphic to $\mathbb{R}$, $\mathbb{C}$ or $\mathbb{H}$ and hence to $\delta_{M_{11}}(M(h,p;\mathbb{R}))$, $\delta_{M_{11}}(M(h,p;\mathbb{C}))$ or $\delta_{M_{11}}(M(h,p;\mathbb{H}))$. Since both these $\mathbb{R}$-J*-algebras satisfy the assumption of lemma 11, the above $\mathbb{R}$-J*-isomorphism, σ, can be chosen to satisfy:

$$\sigma(xy^*z + zy^*x) = \sigma(x)\sigma(y)^*\sigma(z) + \sigma(z)\sigma(y)^*\sigma(x) \quad \text{whenever}$$

$$x,y,z \in \delta_{x_1}(\mathcal{j}).$$

The above considerations than imply that the mapping $\sigma : \mathcal{j} \to M(h,p;\mathbb{K})$ defined by linearity and by $\sigma(\varphi_{ij}(x)) = \varphi_{ij}\,\sigma(x)$ whenever $i = 1,\dots,h$, $j = 1,\dots,p$ and $x \in \delta_{x_1}(\mathcal{j})$, is well posed and it turns out to be a $\mathbb{R}$-J*-isomorphism.

The proof is complete.

<u>LEMMA 38</u>. Let $\mathcal{j}$ be an irreducible $\mathbb{R}$-J*-algebra of height $h \geq 3$ and $\{x_1,\dots,x_h\}$ a subset of $\mathcal{j}$ whose elements are non zero, idempotent, irreducible and pairwise strongly independent. Moreover let $\mathcal{B}$ be a subset of $\psi_{x_1}\psi_{x_2}(\mathcal{j})$ given by lemma 29; If

1) $\mathcal{j} = \Sigma_i\delta_{x_i}(\mathcal{j}) + \Sigma_{rs}\psi_{x_r}\psi_{x_s}(\mathcal{j})$ ($1 \leq i \leq h$ and $1 \leq r < s \leq h$);

2) $|\mathcal{B}| = 2$;

then $\mathcal{j}$ is $\mathbb{R}$-J*-isomorphic to one of the following $\mathbb{R}$-J*-algebras:

$A(2h;\mathbb{R})$, $A(2h;\mathbb{C})$.

PROOF. Let x and y be two elements of $\mathcal{B}$ and $\tilde{x} = 6\{x_1 x_2 x\}$, $\tilde{y} = 6\{x_1 x_2 y\}$. Moreover let $\{c_{ij} : 1 \leqslant i, j \leqslant h\}$ be the subset of $\mathcal{J}$ defined in the proof of lemma 36; let the mapping $\Delta_{ij} : \psi_{x_1}\psi_{x_2}(\mathcal{J}) \to \psi_{x_i}\psi_{x_j}(\mathcal{J})$ for $i \neq j$, be given by $\Delta_{ij}(x) = c_{i1}x^*c_{2j} + c_{j2}x^*c_{1i}$ (we recall that $c_{ij} = c_{ji}$ and that $c_{ii} = \epsilon_i x_i$) and the mappings $\Delta_{1i} : \delta_{x_1}(\mathcal{J}) \to \delta_{x_i}(\mathcal{J})$ be given by $\Delta_{1i}(z) = c_{i1}z^*c_{1i}$.
For $1 \leqslant i, j \leqslant h$ with $i \neq j$, and $x \in \psi_{x_1}\psi_{x_2}(\mathcal{J})$, we have $\Delta_{ij}\Delta_{ji}(x) = c_{11}c^*_{12}xc^*_{12}c_{22} + c_{22}c^*_{21}xc^*_{21}c_{11} = E(x)$. By lemma 30, $\delta_{x_1}(\mathcal{J})$, $\delta_{x_2}(\mathcal{J})$, $\delta_x(\mathcal{J})$, $\delta_{\tilde{x}}(\mathcal{J})$, $\delta_y(\mathcal{J})$ and $\delta_{\tilde{y}}(\mathcal{J})$ are $\mathbb{R}$-J*-algebras $\mathbb{R}$-J*-isomorphic of dimension 1 or 2. The isomorphism $\varphi_{hk} : \delta_k(\mathcal{J}) \to \delta_h(\mathcal{J})$ is given by $\varphi_{hk}(x) = 6\{h\tilde{k}x\}$ whenever $h, k \in \{x_1 = \tilde{x}_2, x_2 = \tilde{x}_1, x, y, \tilde{x}, \tilde{y}\}$ with $h, k, \tilde{h}, \tilde{k}$ distinct elements. Hence we have

$$\mathcal{J} = \Sigma_i \Delta_{1i}(\delta_{x_1}(\mathcal{J})) + \Sigma_{t,\, rs} \Delta_{rs}\varphi_{tx_1}(\delta_{x_1}(\mathcal{J})) \quad (1 \leqslant i \leqslant h;$$
$$1 \leqslant r < s \leqslant h;\ t \in \{x, \tilde{x}, y, \tilde{y}\}).$$

Now let $\{a_1 = x_1, a_p\}$ (where $p = 1$ or 2) be a $\mathbb{R}$-basis of $\delta_{x_1}(\mathcal{J})$ that satisfy (1) of lemma 30, and consider the elements of $A(2h;\mathbb{R})$ $(A(2h;\mathbb{C}))$ given by

$$\sigma(c_{ii}) = M_{2i-1\ 2i} - M_{2i\ 2i-1} \qquad i = 1,\dots,h;$$

$$\sigma(c_{ij}) = \tfrac{1}{2}\left((-1)^{(j-i)/2} - (-1)^{(j-i+1)/2}\right)(M_{2i-1\ 2j-1} - M_{2j-1\ 2i-1} +$$

$$+ M_{2i\ 2j} - M_{2j\ 2i}) + \tfrac{1}{2}\left((-1)^{(j-i)/2} + (-1)^{(j-i+1)/2}\right)(M_{2i-1\ 2j} -$$

$$- M_{2j\ 2i-1} - M_{2i\ 2j-1} + M_{2j-1\ 2i}) = \sigma(c_{ji}) \qquad 1 \leqslant i < j \leqslant h;$$

$$\sigma(x) = M_{13} - M_{31};\ \sigma(\tilde{x}) = \epsilon_2(M_{24} - M_{42});\ \ \sigma(y) = M_{14} - M_{41};$$

$$\sigma(\tilde{y}) = \epsilon_2(M_{32} - M_{23});\ \ \sigma(a_1) = M_{12} - M_{21} \quad (\sigma(a_2) = iM_{12} - iM_{21}).$$

We let the reader prove that $A(2h;\mathbb{R})$ $(A(2h;\mathbb{C}))$ satisfies (1) and that the mapping σ extends to a $\mathbb{R}$-J*-isomorphism of $\mathcal{J}$ onto $A(2h;\mathbb{R})$ $(A(2h;\mathbb{C}))$ such that

i) $\sigma(\varphi_{hc_{11}}(x)) = \varphi_{\sigma(h)\sigma(c_{11})}(\sigma(x))$ whenever $h \in \{x,\tilde{x},y,y\}$ and

$$x = a_1 \quad (x \in \{a_1,a_2\});$$

ii) $\sigma E = E\sigma$;

iii) $\sigma\Delta_{ij} = \Delta_{ij}\sigma$ whenever $1 \leqslant i < j \leqslant h$.

The proof is complete.

LEMMA 39. Let $\mathcal{J}$ be an irreducible $\mathbb{R}$-J*-algebra of height $h \geqslant 2$ and $\{x_1,\ldots,x_h\}$ a subset of $\mathcal{J}$ whose elements are non zero, idempotent, irreducible and pairwise strongly independent. Moreover let $\mathcal{B}$ be a subset of $\psi_{x_1}\psi_{x_2}(\mathcal{J})$ defined by lemma 29. If.

1) $\mathcal{J} \neq \Sigma_i \delta_{x_i}(\mathcal{J}) + \Sigma_{rs}\psi_{x_r}\psi_{x_s}(\mathcal{J})$ $(1 \leqslant i \leqslant h;\ 1 \leqslant r < s \leqslant h)$

2) $|\mathcal{B}| = 2$;

Then $\mathcal{J}$ is $\mathbb{R}$-J*-isomorphic to one of the following $\mathbb{R}$-J*-algebras:

$$A(2h+1;\mathbb{R}),\quad A(2h+1;\mathbb{C}).$$

PROOF. Let us denote by x and y the two elements of $\mathcal{B}$ and let $\tilde{x} = 6\{x_1x_2x\}$, $\tilde{y} = 6\{x_1x_2y\}$ and I_1 be the $\mathbb{R}$-J*-subalgebra of $\mathcal{J}$ given by

$$I_1 = \{y : y \in \mathcal{J} \text{ such that } \psi_{x_1}(y) = y,\ \delta_{x_1}(y) = 0 \text{ and } \psi_{x_1}(y) = 0 \text{ whenever } i = 2,\ldots,h\}.$$

Proceeding as in the proof of lemma 37, we can prove that I_1 decomposes into four $\mathbb{R}$-J*-algebras $I_1 = (I_1 \cap \Gamma_x(\mathcal{J}) \cap \Gamma_y(\mathcal{J})) \cup$ $\cup (I_1 \cap \Gamma_{\tilde{x}}(\mathcal{J}) \cap \Gamma_y(\mathcal{J})) \cup (I_1 \cap \Gamma_x(\mathcal{J}) \cap \Gamma_{\tilde{y}}(\mathcal{J})) \cup (I_1 \cap \Gamma_{\tilde{x}}(\mathcal{J}) \cap \Gamma_{\tilde{y}}(\mathcal{J}))$. Since $I_1 \neq \{0\}$ (see lemma 35), at least one of the above J*-algebras is not zero. Assume $I_1 \cap \Gamma_{\tilde{x}}(\mathcal{J}) \cap \Gamma_{\tilde{y}}(\mathcal{J}) \neq \{0\}$.

Let c be a non zero, idempotent, irreducible element of $I_1 \cap \Gamma_{\tilde{x}}(j) \cap \Gamma_{\tilde{y}}(j)$. c is irreducible in j (see lemma 35) and: $x, y \in \psi_c \psi_{x_r}(j)$ (see lemma 37). Then $6\{x_2 c x\} = cx^*x_2 + x_2 x^* c = t$ and $6\{x_2 c y\} = cy^* x_2 + x_2 y^* c = w$ are elements of $\psi_c \psi_{x_2}(j) \cap I_2$ where I_2 is the set

$$I_2 = \{y : y \in j \text{ such that } \psi_{x_2}(y) = y,\ \delta_{x_2}(y) = 0 \text{ and } \psi_{x_i}(y) = 0 \text{ whenever } i = 1,\dots,h \text{ with } i \neq 2\}.$$

Moreover we have $\delta_t(w) = \delta_w(t) = 0$.
Now, let us prove that $I_2 = \delta_t(j) + \delta_w(j)$. Assume that this is not true, i.e. that there exists a non zero element $z \in I_2$ such that $\delta_t(z) = \delta_w(z) = 0$. Then we would have $0 = (cx^*x_2 + x_2x^*c)z^*(cx^*x_2 + x_2x^*c) = cx^*x_2z^*cx^*x_2 + x_2x^*cz^*x_2x^*c$ and hence $x^*x_2z^*cx = x^*cz^*x_2x = 0$ Thus we would obtain $x^*x_2z^*c = cz^*x_2x^* = 0$ and the similar relations $y^*x_2z^*c = cz^*x_2y^* = 0$. The above relations imply $wz^*t = tz^*w = 0$ and hence (since the height of I_2 is 1) the contradiction $z = 0$.
We have $6\{x_2(x + \tilde{x})t\} = c \in I_1 \cap \Gamma_{\tilde{x}}(j) \cap \Gamma_{\tilde{y}}(j)$. Define $q = 6\{x_2(x + \tilde{x})w\} = (x + \tilde{x})y^*c + cy^*(x + \tilde{x}) = \tilde{x}y^*c + cy^*\tilde{x} = \tilde{y}x^*c + cx^*\tilde{y} \in I_1 \cap \Gamma_x(j) \cap \Gamma_{\tilde{y}}(j)$. Let us denote by $\varphi_{cx_1} : \delta_{x_1}(j) \to \delta_c(j)$ the $\mathbb{R}$-J*-isomorphism defined by $\varphi_{cx_1}(z) = 6\{x_1 \varphi_{xx_1}(z) t\}$, and with $\varphi_{qx_1} : \delta_{x_1}(j) \to \delta_q(j)$ the $\mathbb{R}$-J*-isomorphism given by $\varphi_{qx_1}(z) = 6\{x_1 \varphi_{\tilde{x}x_1}(z) w\}$, and by $\Delta_i : I_1 \to I_i$, $1 \leqslant i \leqslant h$, the $\mathbb{R}$-J*-isomorphism given by $\Delta_i(z) = zc^*_{1i}c_{1i} + c_{i1}c^*_{1i}z$. (Recall that $I_i = \{y . y \in j$ such that $\psi_{x_i}(y) = y$, $\delta_{x_i}(y) = 0$ and $\psi_{x_j}(y) = 0$ whenever $j = 1,\dots,h$ with $j \neq i\}$). It is trivial to prove that

$$j = \Sigma_i \Delta_{i1}(\delta_{x_1}(j)) + \Sigma_{rst} \Delta_{rs} \varphi_{tx_1}(\delta_{x_1}(j)) + \Sigma_{iw} \Delta_i \varphi_{wx_1}(\delta_{x_1}(j))$$

$$(1 \leqslant i \leqslant h;\ 1 \leqslant r < s \leqslant h;\ t \in \{x, y, \tilde{x}, \tilde{y}\};\ w \in \{c, q\}).$$

Preserving the notation established in the previous lemmas, for the $\mathbb{R}$-J*-algebras $A(2h+1;\mathbb{R})$ and $A(2h+1;\mathbb{C})$, we set:

$$\sigma(c) = M_{1\ 2h+1} - M_{2h+1\ 1}, \quad \sigma(q) = M_{2\ 2h+1} - M_{2h+1\ 2}$$

The reader can trivially show that σ extends to a $\mathbb{R}$-J*-isomorphism of $\mathcal{J}$ onto $A(2h+1;\mathbb{R})$ $(A(2h+1;\mathbb{C}))$.

The proof is complete.

So far, we have proved that any finite dimensional, irreducible $\mathbb{R}$-J*-algebra is $\mathbb{R}$-J*-isomorphic to one of the following matrix $\mathbb{R}$-J*-algebras.

1) $M(h,p;\mathbb{R})$ whenever $1 \leqslant h \leqslant p$;

2) $M(h,p;\mathbb{C})$ whenever $2 \leqslant h \leqslant p$;

3) $M(h,p;\mathbb{H})$ whenever $2 \leqslant h \leqslant p$;

4) $S(n;\mathbb{R})$ whenever $3 \leqslant n$,

5) $S(n;\mathbb{C})$ whenever $2 \leqslant n$;

6) $A(n;\mathbb{R})$ whenever $4 \leqslant n$;

7) $A(n;\mathbb{C})$ whenever $4 \leqslant n$;

8) $H(n;\mathbb{C})$ whenever $3 \leqslant n$;

9) $H(n;\mathbb{H})$ whenever $3 \leqslant n$;

10) $H^*(n;\mathbb{H})$ whenever $2 \leqslant n$;

11) $I_{2n,m}$ whenever $1 \leqslant n$, $0 \leqslant m$ but $(m,n) \neq (0,1)$;

12) E_n whenever $3 \leqslant n$.

Before stating our main theorem let us define some more $\mathbb{R}$-J*-isomorphism invariants beside height and $\mathbb{R}$-dimension.

We define

$K(\mathcal{J})$ = Sup{$\dim \psi_a \psi_b(\mathcal{J})$ where a, b are non zero, idempotent, irreducible, strongly independent elements of $\mathcal{J}$};

$H(\mathcal{J})$ = Sup{$\dim \delta_a(\mathcal{J})$: where a is a non zero, idempotent, irreducible element of $\mathcal{J}$};

$S(\mathcal{J})$ = Sup{$|\mathcal{B}_{ab}|$. where a, b are non zero, idempotent, irreducible and strongly independent elements of $\mathcal{J}$ and $\mathcal{B}_{ab}$ is the set defined in lemma 29}.

If we observe that there exist natural $\mathbb{R}$-J*-isomorphisms between:

$\Gamma_{M_{11}}(M(h+1,p+1;\mathbb{R}))$ and $M(h,p;\mathbb{R})$;

$\Gamma_{M_{11}}(M(h+1,p+1;\mathbb{C})$ and $M(h,p;\mathbb{C})$;

$\Gamma_{M_{11}}(M(h+1,p+1;\mathbb{H})$ and $M(h,p;\mathbb{H})$;

$\Gamma_{M_{11}}(S(n+1;\mathbb{R}))$ and $(S(n;\mathbb{R})$;

$\Gamma_{M_{11}}(S(n+1;\mathbb{C}))$ and $S(n;\mathbb{C})$;

$\Gamma_{M_{12}-M_{21}}(A(n+2;\mathbb{R}))$ and $A(n;\mathbb{R})$;

$\Gamma_{M_{12}-M_{21}}(A(n+2;\mathbb{C}))$ and $A(n;\mathbb{C})$;

$\Gamma_{M_{11}}(H(n+1;\mathbb{H}))$ and $H(n;\mathbb{H})$;

$\Gamma_{M_{11}}(H(n+1;\mathbb{C}))$ and $H(n;\mathbb{C})$;

$\Gamma_{iM_{11}}(H^*(n+1;\mathbb{H}))$ and $H^*(n;\mathbb{H})$;

we can prove the following table of the above $\mathbb{R}$-J*-invariants:

j	$\mathbb{R}$-dim j	Hght(j)	K(j)	H(j)	S(j)	Limitations
$M(h,p;\mathbb{R})$	hp	h	2	1	1	$1 \leq h \leq p$
$M(h,p;\mathbb{C})$	2hp	h	4	2	1	$2 \leq h \leq p$
$M(h,p;\mathbb{H})$	4hp	h	8	4	1	$2 \leq h \leq p$
$S(n;\mathbb{R})$	$\frac{n(n+1)}{2}$	$[\frac{n}{2}]$	1	1	0	$3 \leq n$
$S(n;\mathbb{C})$	n(n+1)	$[\frac{n}{2}]$	2	2	0	$2 \leq n$
$A(n;\mathbb{R})$	$\frac{n(n-1)}{2}$	n	4	1	4	$4 \leq n$
$A(n;\mathbb{C})$	n(n-1)	n	8	2	4	$4 \leq n$
$H(n;\mathbb{C})$	n^2	n	2	1	0	$3 \leq n$
$H(n;\mathbb{H})$	2n(n-1)+n	n	4	1	0	$3 \leq n$
$H^*(n;\mathbb{H})$	2n(n-1)+3n	n	4	3	0	$2 \leq n$
$I_{2n,m}$	2n+m	2	2n+m-2	1	n-1	$1 \leq n$, $0 \leq m$, $(m,n) \neq (0,1)$
E_n	2n	2	2n-4	2	$[\frac{n-2}{2}]$	$3 \leq n$

Since E_3 is naturally IR-J*-isomorphic to $S(2;\mathbb{C})$, E_4 to $M(2,2;\mathbb{C})$ and $I_{4,0}$ to $M(2,2;\mathbb{R})$ the above table and the previous lemmas lead to our main theorem:

THEOREM 1. Any non zero irreducible IR-J*-algebra of finite dimension is IR-J*-isomorphic to exactly one of the following irreducible matrix IR-J*-algebras:

1) $M(h,p;\mathbb{R})$ with $1 \leqslant h \leqslant p$;

2) $M(h,p;\mathbb{C})$ with $2 \leqslant h \leqslant p$;

3) $M(h,p;\mathbb{H})$ with $2 \leqslant h \leqslant p$;

4) $S(n;\mathbb{R})$ with $3 \leqslant n$;

5) $S(n;\mathbb{C})$ with $2 \leqslant n$;

6) $A(n;\mathbb{R})$ with $4 \leqslant n$;

7) $A(n;\mathbb{C})$ with $4 \leqslant n$;

8) $H(n;\mathbb{C})$ with $3 \leqslant n$;

9) $H(n;\mathbb{H})$ with $3 \leqslant n$;

10) $H^*(n;\mathbb{H})$ with $2 \leqslant n$;

11) $I_{2n,m}$ with $1 \leqslant n$, $0 \leqslant m$ and $(m,n) \neq (0,1), (0,2)$;

12) E_n with $5 \leqslant n$.

Any non zero, IR-J*-algebra of finite dimension is the IR-J*-isomorphic to the direct sum of a finite number of addend of the above table. This decomposition is unique.

PART 4.

In the present section we obtain some conditions that must be fullfilled by a $\mathbb{R}$-J*-algebra in order to be the $\mathbb{R}$-J*-algebra naturally associated with a $\mathbb{C}$-J*-algebra. Then, using the above conditions and theorem 1, we prove that any non zero, finite dimensional, irreducible $\mathbb{C}$-J*-algebra of height $h \geq 2$ is $\mathbb{R}$-J*-isomorphic to exactly one of the $\mathbb{R}$-J*-algebras of type 2), 5), 7) and 12) (see the table of theorem 1). After proving that any non zero, finite dimensional $\mathbb{C}$-J*-algebra of height 1 is $\mathbb{C}$-J*-isomorphic to $M(1,p;\mathbb{C})$, we prove that the existence of one of the above $\mathbb{R}$-J*-isomorphic assures the existence of a $\mathbb{C}$-J*-isomorphism and hence we obtain our theorem 2.

We terminate this last section concluding that there exists no $\mathbb{C}$-J*-algebra whose unitary disc is one of the two exceptional bounded symmetric domains of the Cartan's classification.

Assume that V and W are two $\mathbb{C}$-Hilbert vector spaces and let $I : V \to V$, $J : W \to W$ be the two $\mathbb{R}$-automorphisms $I(x) = ix$ and $J(x) = ix$. Let now j be a $\mathbb{C}$-J*-algebra of bounded linear operators of V into W. Whenever $a \in j$ we have:

1) $J_a = aI$;

2) $(Ja)^* = - a^*I$.

<u>LEMMA 40</u>. Let j be a non zero $\mathbb{C}$-J*-algebra of height $h \geq 2$ and let $a \in j$ be a non zero, idempotent, irreducible element. Then the $\mathbb{R}$-dimension of $\delta_a(j)$ is exactly 2.

<u>PROOF</u>. Assume that a is a non zero, idempotent, irreducible element of j. Ja, and a are $\mathbb{R}$-linearly independent elements of j and since $\delta_a(Ja) = aa^*Jaa^*a = Jaa^*aa^*a = Ja$, we have $\dim_{\mathbb{R}}\delta_a(j) \geq 2$. $\delta_a(j)$ is a $\mathbb{R}$-J*-subalgebra of j which satisfies the assumption of lemma 11. Since $a^*(Ja) + (Ja)^*a = a^*Ja - a^*Ja = 0$ and $(Ja)a^* + a(Ja)^* = 0$, the assumption that $\dim_{\mathbb{R}}(\delta_a(j) > 2$ implies that there exists a non zero,

idempotent, irreducible element $b \in \delta_a(j)$ such that: $a^*b + b^*a = ab^* + ba^* = 0$; $(b^*a - a^*b) = 0$; $(ba^* - ab^*) = 0$. These relations imply the contradiction $ba^* = ab^* = b^*a = a^*b = 0$.

LEMMA 41. Let j be a $\mathbb{C}$-J*-algebra of height 1. j is $\mathbb{C}$-J*-isomorphic to $M(1,p;\mathbb{C})$ where $p = \dim_{\mathbb{C}}(j)$.

PROOF. The same as in lemma 34.

LEMMA 42. Let j be a non zero, irreducible, finite dimensional $\mathbb{C}$-J*-algebra. j is $\mathbb{R}$-J*-isomorphic to exactly one of the following $\mathbb{C}$-J*-algebras:

1) $M(h,p;\mathbb{C})$ with $1 \leq h \leq p$;

2) $A(n;\mathbb{C})$ with $4 \leq n$;

3) $S(n;\mathbb{C})$ with $2 \leq n$;

4) E_n with $5 \leq n$.

PROOF. It is a consequence of theorem 1, lemmas 40 and 41 and the table of invariants that precedes theorem 1.

Let j be a $\mathbb{C}$-J*-algebra of complex matrices. The set $j' = \{\bar{a} : a \in j\}$ is a $\mathbb{C}$-J*-algebra $\mathbb{R}$-J*-isomorphic to j.

LEMMA 43. Let j be a $\mathbb{C}$-J*-algebra $\mathbb{R}$-J*-isomorphic to one of the $\mathbb{C}$-J*-algebras:

1) $M(h,p;\mathbb{C})$ with $1 \leq h \leq p$;

2) $A(n;\mathbb{C})$ with $4 \leq n$;

3) $S(n;\mathbb{C})$ with $2 \leq n$;

4) E_n with $5 \leq n$.

Hence j is $\mathbb{C}$-J*-isomorphic to the same matrix algebra.

PROOF. Let us consider separately the four cases.

Assume $\sigma : M(h,p;\mathbb{C}) \to \mathcal{J}$ is any $\mathbb{R}$-J*-isomorphism. By lemma 40 and lemma 11, we have only two possibilities:

i) $\sigma(iM_{11}) = i\sigma(M_{11})$;

ii) $\sigma(iM_{11}) = -i\sigma(M_{11})$.

Using, if necessary, the $\mathbb{R}$-J*-isomorphism $\varphi : M(h,p;\mathbb{C}) \to M(h,p;\mathbb{C})$ defined by $\varphi(x) = \bar{x}$, we may assume that i) is verified. Hence, whenever $x \in \delta_{M_{11}}(M(h,p;\mathbb{C}))$ and $\lambda \in \mathbb{C}$, we have $\sigma(\lambda x) = \lambda\sigma(x)$. Using the notation established in lemma 37, whenever $x \in \delta_{M_{11}}(M(h,p;\mathbb{C}))$, $i = 1,\ldots,h$, $j = 1,\ldots,p$ and $\lambda \in \mathbb{C}$, we have

$$\sigma(\lambda x) = \sigma(\varphi_{ij}\varphi_{ij}^{-1}(\lambda x)) = \sigma(\varphi_{ij}(\bar{\lambda}\varphi_{ij}^{-1}(x)) = \varphi_{ij}(\bar{\lambda}\sigma\varphi_{ij}^{-1}(x)) =$$

$$= \lambda\varphi_{ij}\varphi_{ij}^{-1}(x) = \lambda\sigma(x)$$

and hence σ is a $\mathbb{C}$-linear isomorphism.

The other cases are proved in the same way.

<u>THEOREM 2</u>. Any non zero, irreducible $\mathbb{C}$-J*-algebra of finite dimension is $\mathbb{C}$-J*-isomorphic to exactly one of the following matrix $\mathbb{C}$-J*-algebras

1) $M(h,p;\mathbb{C})$ with $1 \leqslant h \leqslant p$;

2) $A(n;\mathbb{C})$ with $4 \leqslant n$;

3) $S(n;\mathbb{C})$ with $2 \leqslant n$;

4) E_n with $5 \leqslant n$.

Any non zero, finite dimensional $\mathbb{C}$-J*-algebra is $\mathbb{C}$-J*-isomorphic to a direct sum of a finite number of addend of the above table. This decomposition is unique.

In a paper published in 1973, [1], L.A. Harris proved that the unitary disc of a finite dimensional $\mathbb{C}$-J*-algebra is a finite dimensional bounded symmetric domain. Moreover, he showed that any domain of the classes I),...,IV) in Cartan's classification can be obtained as a disc of a J*-algebra. Since the canonical realization of Cartan's domains of type I),...,IV) are the unitary discs of the matrix ℂ-J*-algebras enumerated in theorem 2, we obtain:

COROLLARY 7. Any non zero, finite dimensional ℂ-J*-algebra $\mathcal{J}$ has a unitary disc decomposable in Cartan's factors of type I),...,IV).

COROLLARY 8. The two exceptional domains cannot be obtained as unitary disc of any ℂ-J*-algebra.

We have obtained here an independent proof of the result of O. Loss and K. McCrimmon [3].

REFERENCES

[1] L.A. HARRIS "Bounded symmetric homogeneous domains in infinite dimensional spaces", Lectures Notes in Mathematics n. 364, 13-40; Springer Verlag (1973).

[2] O. LOOS-K. MCCRIMMON "Speciality of Jordan triple sistems", Comm. Alg. 5(1977), n. 10, 1057-1082.

[3] E. CARTAN "Sur les domaines bornés homogènes de l'espace de n variables complexes", Abh. Math. Sem. Univ. Hamburg 11 (1935), 116-162.

[4] E. VESENTINI "Alcuni aspetti della geometria dei domini limitati", Rend. Sem. Mat. Fis. Milano, 50 (1980), 109-115.

JOHN H. RAWNSLEY

F-STRUCTURES, F-TWISTOR SPACES AND HARMONIC MAPS

1. INTRODUCTION

In their paper [5] J. Eells and J.C. Wood fibre non-holomorphically certain partial flag manifolds $H_{r,s}$ over the complex projective space $\mathbb{C}P^{r+s}$. They observe that horizontal holomorphic maps into $H_{r,s}$ compose with the projection map to give harmonic maps into $\mathbb{C}P^{r+s}$. These notes grew out of an attempt to understand this result. The bundles $H_{r,s}$ are examples of (metric) twistor spaces studied by N. O'Brian and the author in [10], and it became clear that any horizontal holomorphic map into a metric twistor space gives a harmonic map into the base of the bundle.

The situation was clarified by J. Eells and S. Salamon [4] who observed that if a (non-integrable) almost complex structure J_2 was defined on a twistor space by reversing the complex structure on the fibre then any J_2-holomorphic map into the twistor space projects to a harmonic map into the base. Those which are horizontal and holomorphic are J_2-holomorphic, and the extra information involved in being horizontal manifests itself in isotropy properties of the projected map.

Eells and Salamon also pointed out that for their results it was only necessary to have partial complex structures, so we introduced the notion of f-twistor spaces and found that our results for twistor spaces work just as well in this context. This also led us to examine manifolds with f-structures [12], and obtain some mild generalizations of results of Lichnerowicz [9] on the harmonicity of maps between Hermitian manifolds.

For the theory of harmonic maps and a comprehensive bibliography see the survey article by J. Eells and L. Lemaire [3].

I thank F. Burstall, J. Eells and S. Salamon for keeping me informed of their work, from which I have benefitted enormously. I have also enjoyed useful discussions with J. Glazebrook, M. Reid, B. Smyth and J. C. Wood.

2. F-STRUCTURES

Let (V,h_o) be a real inner-product space of dimension n and k an integer with $0 < 2k \leq n$. Denote by $F_k(V,h_o)$ the set of skew-symmetric endomorphisms F of V of rank $2k$ satisfying

$$F^3 + F = 0$$

F is diagonalizable on the complexification $V^{\mathbb{C}}$ of V with eigenvalues i, 0, $-i$, and corresponding eigenspaces $V^{+,F}$, $V^{0,F}$, $V^{-,F}$. We have

$$(2.1)\quad V^{\mathbb{C}} = V^{+,F} + V^{0,F} + V^{-,F}, \quad V^{-,F} = \overline{V^{+,F}}$$

If $Q^{+,F}$, $Q^{0,F}$, $Q^{-,F}$ denote the corresponding projections then

$$1 = Q^{+,F} + Q^{0,F} + Q^{-,F}, \; Q^{-,F} = \overline{Q^{+,F}}, \; F = i(Q^{+,F} - Q^{-,F}).$$

Let h_o denote also the $\mathbb{C}$-bilinear extension of h_o to $V^{\mathbb{C}}$, then

$$<v,w> = h_o(v,\bar{w})$$

is a Hermitian structure on $V^{\mathbb{C}}$ and the decomposition (2.1) is orthogonal.

<u>Proposition 2.1</u> $V^{+,F}$ is an h_o-isotropic k-dimensional subspace of $V^{\mathbb{C}}$ and $F \leftrightarrow V^{+,F}$ sets up a bijection between elements of $F_k(V,h_o)$ and k-dimensional h_o-isotropic subspaces of $V^{\mathbb{C}}$.

<u>Proof</u> Isotropy follows because eigenspaces $V^{+,F}$, $V^{-,F}$ of different eigenvalues are Hermitian orthogonal. Conversely, given an isotropic subspace W, then W and $\bar{W}$ are orthogonal, so we can define F on

$V^{\mathbb{C}}$ as i on W; $-i$ on $\overline{W}$ and 0 on $(W+\overline{W})^{\perp}$. It is simple to check that F is in fact real, so is in $F_k(V,h_o)$.

An alternative way to view elements of $F_k(V,h_o)$ is to observe that F is invertible on its image, and so defines a Hermitian structure there (on restricting the inner-product of V) since $(F|_{ImF})^2 = -1_{ImF}$ Conversely given a 2k-dimensional subspace of V on which is defined a complex structure compatible with the restriction of the inner-product we can obviously define an f-structure of rank $2k$ on V. It is then clear that the group of real orthogonal endomorphisms of V which commute with a given F is isomorphic to $U(ImF) \times O((ImF)^{\perp})$.

If E is a real smooth vector bundle over a manifold N with a fibre metric h, let $F_k(E,h)$ be the bundle whose fibre at x is $F_k(E_x,h_x)$. A smooth section F of $F_k(E,h)$ will be called an f-structure of rank $2k$ on E. Clearly, an f-structure on E of rank $2k$ corresponds with a reduction of the orthogonal frame bundle of E from an $O(r)$ bundle to a $U(k) \times O(r-2k)$ bundle. The associated characteristic classes are examined in section 11. If F is an f-structure on E then the complexification $E^{\mathbb{C}}$ splits into eigenbundles:

$$E^{\mathbb{C}} = E^+ + E^0 + E^-$$

where

$$(E^+)_N = (E_x)^{+,F_x}, \quad x \in N$$

and so on. We have the analogue of Proposition 2.1 for bundles:

<u>Proposition 2.2</u> There is a bijection between the f-structures F of rank $2k$ on E and the h-isotropic subbundles E^+ of $E^{\mathbb{C}}$ of rank k given by $E^+ = Ker(F-i)$.

If ∇ is a metric connection in E then we can covariantly differentiate an f-structure F in E. We say F is parallel if $\nabla F = 0$.

Lemma 2.3 F is a parallel f-structure in E for the metric connection ∇ if and only if

$$\nabla_X C^\infty(E^+) \subset C^\infty(E^+) \quad \text{for all vector fields } X.$$

Proof Left to the reader.

In the case (N,h) is a Riemannian manifold, an f-structure on TN will be called an f-structure on N, and we shall call (N,h,F) an almost f-manifold. f-structures on a manifold are generalizations of almost Hermitian structures (which correspond with dim N = 2k) and include CR-structures (dim N = 2k+1). They were introduced by Yano [12].

We shall define a morphism of f-manifolds to be a smooth map whose differential intertwines the f-structures

$$d\varphi \circ F^M = F^N \circ d\varphi, \quad \varphi : M \to N.$$

Equivalently

$$d\varphi(T^oM) \subset T^oN, \quad d\varphi(T^\pm M) \subset T^\pm N.$$

In the case the domain is actually almost Hermitian we shall say φ is f-holomorphic, or, if we want to emphasize the f-structure F of the codomain, F-holomorphic. Thus φ is F-holomorphic if and only if

$$d\varphi(T'M) \subset T^+N$$

where we denote T^+M by $T'M$ and T^-M by $T''M$ when the f-structure is an almost Hermitian structure.

The analogue for f-structures of the Kaehler condition for an almos Hermitian structure is that F be parallel. This is very strong (it implies the integrability of all the distributions T^oN, T^+N, $(T^oN)^\perp$, for instance). Instead we consider a weaker condition which we shall call condition A for want of a better name. We say F satisfies condition A if

(A) $\nabla_X C^\infty(T^+N) \subset C^\infty(T^+N), \quad X \in C^\infty(T^-N).$

For an almost Hermitian structure this is equivalent to

$$(d\omega)^{1,2} = 0$$

where ω is the Kaehler 2-form. An almost Hermitian manifold satisfying this condition is said to be (1,2)-symplectic.

Proposition 2.4 If $F \in F_k(V,h_o)$ has associated decomposition

$$V^{+,F} + V^{0,F} + V^{-,F}$$

then

$$\{v \in V^{\mathbb{C}} \quad h(v,w) = 0 \ , \ \forall w \in V^{+,F}\} = V^{0,F} + V^{+,F}$$

Proof Elementary.

Proposition 2.5 If F is an f-structure on a bundle E and ∇ is a metric connection on E then $\nabla_X F$ maps each eigen-bundle of F to the sum of the other two.

Proof Elementary.

Proposition 2.6 An f-structure F on a Riemannian manifold (N,h) satisfies condition A if and only if

$$(\nabla_X F)(T^+N) = 0 \ , \quad \forall X \in T^-N.$$

Proof If Y is in T^+N then

$$(\nabla_X F)Y = (\nabla_X (F-i))Y = -(F-i)\nabla_X Y$$

and the result follows.

Our first application to the theory of harmonic maps is the following generalization of the result (see [9]) that holomorphic maps of Kaehler manifolds are harmonic.

Theorem 2.7 Let (M,g,J) be almost Hermitian with coclosed Kaehler form and let (N,h,F) be a Riemannian manifold whose f-structure satisfies condition A. Then every f-holomorphic map $\varphi : M \to N$ is harmonic

Proof The tension field τ_φ is given by

$$\tau_\varphi = \sum_i \nabla d\varphi(X_i, X_i)$$

for any orthonormal frame $(X_1, \ldots, X_{2m})$ on M. We can take a frame of the form $(Y_1,\ldots,Y_m, JY_1,\ldots,JY_m)$, then

$$\tau_\varphi = \sum_j \nabla d\varphi(Y_j, Y_j) + \nabla d\varphi(JY_j, JY_j)$$

$$= \sum_j \nabla d\varphi(\bar{Z}_j, Z_j)$$

where $Z_j = Y_j - JY_j \in T'M$. But $d^*\omega^M = 0$ is equivalent to

$$\sum_j (\nabla_{X_j} J)(X_j) = 0$$

which is equivalent to

$$\sum_j \nabla_{\bar{Z}_j} Z_j \in T'M .$$

Then

$$\tau_\varphi = \sum_j (\varphi^{-1}\nabla)_{\bar{Z}_j} \varphi_* Z_j - \varphi_* \nabla_{\bar{Z}_j} Z_j .$$

But $\varphi_* Z \in \varphi^{-1}T^+N$ since φ is f-holomorphic, and if F satisfies condition A, $(\varphi^{-1}\nabla)_{\overline{Z}_j}\varphi_*Z \in \varphi^{-1}T^+N$. Hence

$$\tau_\varphi \in \varphi^{-1}T^+N.$$

But τ_φ is real, so

$$\tau_\varphi \in \varphi^{-1}T^+N \cap \varphi^{-1}T^-N = 0 ,$$

and hence φ is harmonic.

If (M,g,J) is (1,2)-symplectic, then, in the notation of the above proof, $\nabla_{\overline{Z}_j} Z_k$ is (1,0) so

$$(\nabla d\varphi)(\overline{Z}_j, Z_k) = (\varphi^{-1}\nabla)_{\overline{Z}_j} d\varphi(Z_k) - d\varphi\nabla_{\overline{Z}_j} Z_k$$

is in $\varphi^{-1}T^+N$ when F satisfies condition A. Interchanging $\overline{Z}_j$ and Z_k shows it is in $\varphi^{-1}T^-N$ and hence vanishes. Thus we have shown

<u>Theorem 2.8</u> If $\varphi : M \to N$ is an f-holomorphic map of a (1,2)-symplectic manifold M into N whose f-structure satisfies condition A then the (1,1) part $(\nabla d\varphi)^{1,1}$ of the second fundamental form of φ vanishes.

<u>Remark 2.9</u> The condition $(\nabla d\varphi)^{1,1} = 0$ is quite interesting. It is weaker than holomorphic and stronger than harmonic in general. From the composition law

$$\nabla d(\varphi \circ \psi) = \varphi_* \nabla d\psi + \psi^* \nabla d\varphi$$

it follows that $(\nabla d\varphi)^{1,1} = 0$ and ψ a holomorphic map of (1,2)-symplectic manifolds, then $(\nabla d(\varphi \circ \psi))^{1,1} = 0$. In particular φ when restricted to any almost complex submanifold will also satisfy the condition. If the domain has one complex dimension then $(\nabla d\varphi)^{1,1} = \tau_\varphi \otimes g$ so the condition is the same as being harmonic. Conversely, suppose φ has the property of being harmonic when restricted to any 1-dimensional complex submanifold then applying the composition formula easily leads to the conclusion $(\nabla d\varphi)^{1,1} = 0$. Thus the vanishing of $(\nabla d\varphi)^{1,1}$ is equivalent to φ being harmonic when restricted to complex curves.

The decomposition

$$TN^{\mathbb{C}} = T^{+}N + T^{o}N + T^{-}N$$

induces a corresponding decomposition of the differential of a map into N. In the case the domain is an almost Hermitian manifold (M,g,J) and (N,h,F) is an almost f-manifold we let

$$\omega(X,Y) = g(X,JY)$$

and

$$\sigma(X,Y) = h(X,FY)$$

be the Kaehler 2-forms on M and N respectively. Let $J^{\pm}$ be the projections onto the $\pm i$ eigenspaces of J and $Q^{\pm,F}$, $Q^{o,F}$ the projections onto the $\pm i$, 0 eigenspaces of F. If $\varphi : M \to N$ is a smooth map we define

$$d^{+}\varphi = (\varphi^{-1}Q^{+,F}) \circ d\varphi \circ J^{+};$$

$$d^{-}\varphi = (\varphi^{-1}Q^{+,F}) \circ d\varphi \circ J^{-};$$

$$d^{o}\varphi = (\varphi^{-1}Q^{o,F}) \circ d\varphi .$$

If we set

$$e^{\pm}(\varphi) = |d^{\pm}\varphi|^{2}, \quad e^{o}(\varphi) = \tfrac{1}{2}|d^{o}\varphi|^{2}$$

then the energy density $e(\varphi)$ of φ decomposes

$$e(\varphi) = e^{+}(\varphi) + e^{-}(\varphi) + e^{o}(\varphi) .$$

If $<\,,\,>$ denotes the fibre inner-product which g induces on 2 forms on M we have the

<u>Proposition 2.10</u>

$$e^{+}(\varphi) - e^{-}(\varphi) = <\varphi^{*}\sigma, \omega>$$

<u>Proof</u> Take an orthonormal frame for TM of the form $(X_1,\ldots,X_m, JX_1,\ldots,JX_m)$ and put

$$Z_j = \frac{1}{\sqrt{2}}(X_j - iJX_j),$$

so that $Z_1,\ldots,Z_m$ is an orthonormal frame for $T'M$. Then

$$e^{+}(\varphi) = |d^{+}\varphi|^{2} = \sum_j \varphi^{-1}h(d^{+}\varphi(Z_j), \overline{d^{+}\varphi(Z_j)})$$

$$= \sum_j \varphi^{-1}h(\varphi^{-1}Q^{+,F}d\varphi(Z_j), d\varphi(\overline{Z}_j)).$$

Likewise

$$e^{-}(\varphi) = \sum_j \varphi^{-1}h((\varphi^{-1}Q^{-,F})d\varphi(Z_j), d\varphi(\overline{Z}_j)).$$

But

$$F = iQ^{+,F} - iQ^{-,F}$$

so

$$ie^{+}(\varphi) - ie^{-}(\varphi) = \sum_j \varphi^{-1}h(\varphi^{-1}Fd\varphi(Z_j), d\varphi(\overline{Z}_j))$$

$$= -\sum_j (\varphi^{*}\sigma)(Z_j, \overline{Z}_j).$$

On the other hand $\omega(X_i, X_j) = \omega(JX_i, JX_j) = 0$ so

$$<\varphi^*\sigma,\omega> = \sum_{i,j}(\varphi^*\sigma)(X_i, JX_j)\,\omega(X_i, JX_j)$$

$$= -\sum_j(\varphi^*\sigma)(X_j, JX_j)$$

$$= i\sum_j (\varphi^*\sigma)(Z_j, \overline{Z}_j).$$

Hence

$$e^+(\varphi) - e^-(\varphi) = <\varphi^*\sigma, \omega>$$

as required.

If we define, when M is compact,

$$K(\varphi) = \int_M \{e^+(\varphi) - e^-(\varphi)\}\ \text{vol}$$

then

$$K(\varphi) = (\varphi^*\sigma, \omega).$$

Consider now a curve φ_t of maps and let

$$X_t = d/dt\varphi_t$$

be its derivative. Then we have the homotopy formula for any p-form θ :

$$d/dt\varphi_t^*\theta = (X_t \lrcorner \varphi_t^{-1} d\theta) \circ \Lambda^p d\varphi_t + d\{(X_t \lrcorner \varphi_t^{-1}\theta) \circ \Lambda^{p-1} d\varphi_t\}$$

and so

$$d/dt\ K(\varphi_t) = ((X_t \lrcorner \varphi_t^{-1} d\sigma) \circ \Lambda^2 d\varphi_t, \omega) + ((X_t \lrcorner \varphi_t^{-1}\sigma) \circ d\varphi_t, d^*\omega).$$

Thus we obtain the result:

<u>Theorem 2.11</u> If $d\sigma = 0$ and $d^*\omega = 0$ then $K(\varphi)$ is a homotopy invariant

3. THE COMPLEX MANIFOLD $F_k(V,h_o)$

As in the previous section we let V be a real inner-product space and denote by $O(V)$ its orthogonal group with Lie algebra $o(V)$ of skew-symmetric transformations. $F_k(V,h_o)$ denotes the set of rank $2k$ f-structures on V, which is a subset of the Lie algebra $o(V)$. In fact it forms a single conjugacy class under the adjoint action of $O(V)$. As such it has an invariant complex structure which may be described as follows. Recall that we can identify F with the isotropic subspace $V^{+,F}$ of $V^{\mathbb{C}}$ which it defines. This gives a map

$$\rho_o : F_k(V,h_o) \to G_k(V^{\mathbb{C}})$$

to the Grassmannian of k-dimensional subspaces of $V^{\mathbb{C}}$ which is $O(V)$ equivariant and whose image is the quadric

$$\{W \in G_k(V^{\mathbb{C}}) : h_o(W,W) = 0\}.$$

The latter is clearly a complex submanifold of $G_k(V^{\mathbb{C}})$ and makes $F_k(V,h_o)$ into a complex manifold as claimed. We shall describe this complex structure in another way below which will be useful in later sections.

For F in $F_k(V,h_o)$ let the stability group of the $O(V)$ action and its Lie algebra be

$$H^F = \{g \in O(V) : gFg^{-1} = F\};$$

$$h^F = \{A \in o(V) : [A,F] = 0\}.$$

More generally, let G be any compact Lie group, g its Lie algebra and

$C \subset g$ a conjugacy class. Then C has an invariant complex structure which may be defined as follows For $\xi \in C$ the stability subgroup and its Lie algebra are

$$H^{\xi} = \{ g \in G \quad Ad\, g\, \xi = \xi \},$$

$$h^{\xi} = \{ \eta \in g \quad [\eta, \xi] = 0 \},$$

so

$$C \cong G/H^{\xi}, \quad T_{\xi}C \cong g/h^{\xi}.$$

If

$$ad\,\xi(\eta) = [\eta, \xi]$$

then, fixing a G-invariant inner-product on g, $ad\,\xi$ is skew-symmetric with h^{ξ} as kernel. It follows that the image m^{ξ} of $ad\,\xi$ is a complement for h^{ξ}. Thus we have

<u>Proposition 3.1</u> $m^{\xi} = [\xi, g]$ is a complement for h^{ξ} in g, and

$$T_{\xi}C \cong m^{\xi}.$$

Hence we have to define an H^{ξ}-invariant splitting of the complexification $(m^{\xi})^{\mathbb{C}}$ of m^{ξ} into (1,0) and (0,1) types. But on $(m^{\xi})^{\mathbb{C}}$ $ad\,\xi$ is real, skew-symmetric with non-zero eigenvalues, so we can let $m^{+,\xi}$ be the sum of all eigenspaces whose eigenvalues are above the real axis and $m^{-,\xi}$ the sum over those below. Since $ad\,\xi$ is real, the eigen values occur in conjugate pairs so we have

$$(m^{\xi})^{\mathbb{C}} = m^{+,\xi} + m^{-,\xi}, \quad m^{-,\xi} = \overline{m^{+,\xi}}, \quad m^{+,\xi} \cap m^{-,\xi} = 0.$$

$m^{+,\xi}$ is H-invariant so gives C a unique invariant almost complex structure whose (1,0) tangent space at ξ coincides with $m^{+,\xi}$. That this almost complex structure is integrable follows from the easily verified identity

$$[m^{+,\xi}, m^{+,\xi}] \subset m^{+,\xi}.$$

We summarize this in:

<u>Proposition 3.2</u> A conjugacy class C in g has a G-invariant complex structure whose (1,0) tangent space at $\xi \in C$ is the sum of the eigenspaces of $\mathrm{ad}\,\xi$ whose eigenvalue is above the real axis.

In the case of $F_k(V,h_o) \subset o(V)$, it is easy to verify that ad F can only have eigenvalues $\pm 2i$, $\pm i$, 0 $m^{+,F}$ in this case can be described as follows: We can identify $o(V)$ with $\Lambda^2 V$ by

$$v \wedge w(u) = h_o(v,u)w - h_o(w,u)v.$$

On $V^{\mathbb{C}}$ F has eigenvalues i, $-i$, 0 with eigenspaces $V^{+,F}$, $V^{-,F}$, $V^{0,F}$. Then

$$\Lambda^2 V^{\mathbb{C}} = \Lambda^2 V^{+,F} + V^{+,F} \otimes V^{0,F} + \Lambda^2 V^{0,F}$$

$$+ V^{+,F} \otimes V^{-,F} + V^{-,F} \otimes V^{0,F} + \Lambda^2 V^{-,F}.$$

Examination of eigenvalues shows that under the isomorphism of $\Lambda^2 V^{\mathbb{C}}$ with $o(V)^{\mathbb{C}}$ we have

$$m^{+,F} = \Lambda^2 V^{+,F} + V^{+,F} \otimes V^{0,F}$$

(3.1)

$$(h^F)^{\mathbb{C}} = \Lambda^2 V^{0,F} + V^{+,F} \otimes V^{0,F}$$

We can also characterize $m^{+,F}$ as follows:

<u>Proposition 3.3</u>

$$m^{+,F} = \{A \in o(V)^{\mathbb{C}} : AV^{+,F} = 0,\ AV^{0,F} \subset V^{+,F},\ AV^{-,F} \subset V^{0,F} + V^{+,F}\}.$$

Proof It is clear by examining eigenvalues that $m^{+,F}$ is contained in the right hand side Thus given A in $o(V)$, assume it satisfies

$$AV^{+,F} = 0, AV^{0,F} \subset V^{+,F}, AV^{-,F} \subset V^{0,F} + V^{+,F}.$$

Write $A = A^{+} + A^{0} + A^{-}$ with $A \in m^{\pm,F}$, $A^{0} \in (h^{F})^{\mathbb{C}}$ then

$$AV^{+,F} = A^{+}V^{+,F} + A^{0}V^{+,F} + A^{-}V^{+,F} = 0$$

implies

$$A^{+}V^{+,F} = 0, \quad A^{0}V^{+,F} = 0, \quad A^{-}V^{+,F} = 0.$$

Similarly

$$AV^{0,F} = A^{+}V^{0,F} + A^{0}V^{0,F} + A^{-}V^{0,F} \subset V^{+,F}$$

implies

$$A^{0}V^{0,F} = 0, \quad A^{-}V^{0,F} = 0.$$

Finally

$$AV^{-,F} = A^{+}V^{-,F} + A^{0}V^{-,F} + A^{-}V^{-,F} \subset V^{0,F} + V^{+,F}$$

implies

$$A^{0}V^{-,F} = 0, \quad A^{-}V^{-,F} = 0.$$

Thus $A^{0} = 0$, $A^{-} = 0$ and we have proven the proposition.

4 THE F-TWISTOR SPACES $F_k(N,h)$

Let (N,h) be a Riemannian manifold and V a real inner product space of the same dimension. Let $O(N,h)$ be the frame bundle of (N,h) with fibre at x consisting of all linear isometries

$$p : V \to T_xN.$$

This is a principal $O(V)$-bundle with $O(V)$ acting by composition on the right. Let R_g denote the right translation of $O(N,h)$ by g in $O(V)$ and for A in $o(V)$ let $\tilde{A}$ be the vertical vector field

$$\tilde{A}_p = d/dt|_o \, p.\exp tA.$$

The soldering 1-form θ on $O(N,h)$ is the V-valued 1-form given by

$$\theta_p(X) = p^{-1} d\pi_o X$$

where

$$\pi_o : O(N,h) \longrightarrow N$$

is the bundle projection. Obviously

$$\theta(\tilde{A}) = 0, \quad R_g{}^*\theta = g^{-1}\theta.$$

A connection in $O(N,h)$ (or a metric connection in TN) is an $o(V)$-valued 1-form α on $O(N,h)$ with

$$\alpha(\tilde{A}) = A, \quad R_g{}^*\alpha = \mathrm{Ad}g^{-1}\alpha.$$

α determines a horizontal distribution H^{α} on $O(N,h)$ given by

$$H_p^{\alpha} = \mathrm{Ker}\alpha_p : T_pO(N,h) \longrightarrow o(V).$$

Then

$$d\pi_o \cdot H_p^{\alpha} \longrightarrow T_{\pi_o p}N$$

is an isomorphism whose inverse is called the horizontal lift at p and will be denoted by H_p^{α}.

Let $F_k(N,h)$ be the bundle whose fibre at x is $F_k(T_xN,h_x)$. This is clearly the same as $O(N,h)\times_{O(V)}F_k(V,h_o)$. Fixing F_o in $F_k(V,h_o)$, then since

$$F_k(V,h_o) = O(V)/H^{F_o},$$

$F_k(N,h)$ can also be identified with

$$O(N,h)/H^{F_o}$$

and so we have a principal H^{F_o}-bundle

$$\pi_1 : O(N,h) \longrightarrow F_k(N,h), \quad p \longrightarrow pF_op^{-1}.$$

Let

$$\pi : F_k(N,h) \longrightarrow N$$

be the bundle projection Consider the pull-back $\pi^{-1}TN$ to $F_k(N,h)$. To simplify notation, set

$$Z = F_k(N,h), \quad E = \pi^{-1}TN.$$

At $F \in Z$, if $x = \pi F$ then $F \in \mathrm{End}\, T_xN$ and so we get a canonica section $\mathcal{F}$ of End E given by

$$\mathcal{F}_F = F.$$

Thus E has a canonical f-structure and $E^{\mathbb{C}}$ splits into eigenbundles

$$E^{\mathbb{C}} = E^+ + E^0 + E^-.$$

Since $TN = O(N,h)x_{O(V)}V$, we have

$$E = O(N,h)x_{H}F_O V$$

and so

$$E^{\pm} = O(n,h)x_{H}F_O V^{\pm,F_O}, \quad E^0 = O(N,h)x_{H}F_O V^{0,F_O}.$$

Let $\mathcal{V}$ be the vertical tangent bundle on Z. Thus we have an exact sequence

$$0 \longrightarrow \mathcal{V} \longrightarrow TZ \xrightarrow{d\pi} E \longrightarrow 0. \tag{4.1}$$

Each fibre of π is of the form $F_k(T_xN,h_x) \subset End(T_xN)$ and in fact

$$\pi^{-1}(x) = F_k(T_xN,h_x) \subset o(T_xN)$$

and so if $x = \pi F$ then

$$\mathcal{V}_F = T_F F_k(T_xN,h_x) \cong [F,o(T_xN)] \subset o(T_xN)$$

using Proposition 3.1. Thus we have a natural embedding

$$\mathcal{V} \longrightarrow End\ E$$

as the image of the canonical section $ad\mathcal{F} \in C^{\infty}(End\ E)$. Moreover each fibre has the invariant complex structure described in section 3 which splits $V^{\mathbb{C}}$ as

$$V^{\mathbb{C}} = V' \oplus V''.$$

We have

$$V = O(N,h) \times_{H^{F_o}} m^{F_o}$$

and

$$V' = O(N,h) \times_{H^{F_o}} m^{+,F_o}, \qquad V'' = O(N,h) \times_{H^{F_o}} m^{-,F_o}.$$

Consider a metric connection α on $O(N,h)$ then $\pi^{-1}\nabla^{\alpha}$ is a co-variant derivative in E and induces one also in End E Thus we can form $\pi^{-1}\nabla^{\alpha}F$ which is a 1-form on Z with values in End E.

Proposition 4.1 $\pi^{-1}\nabla^{\alpha}F$ takes values in $V \subset$ End E.

We shall prove a stronger result in a moment. The horizontal distribution H^{α} on $O(N,h)$ is $O(V)$- and hence H^{F_o}-invariant, so projects to $Z = O(N,h)/H^{F_o}$ to give a horizontal distribution H^{α} on Z. Thus

$$TZ = V + H^{\alpha}$$

is a direct sum and we have a projection

$$P^{\alpha} \quad TZ \longrightarrow V$$

which may be viewed as a V-valued 1-form on Z.

Proposition 4.2 As V-valued 1-forms $\pi^{-1}\nabla^{\alpha}F$ and P^{α} are related by

$$\pi^{-1}\nabla^{\alpha}F = [P^{\alpha}, F]$$

<u>Proof</u> E is associated to $O(N,h)$ with fibre V so End E is associated with fibre End V. F then corresponds with the constant function

$$F(p) = F_o$$

on $O(N,h)$. Then $\pi^{-1}\nabla^{\alpha}F$ is given by the End V-valued 1-form

$$dF + [\alpha, F] = [\alpha, F]$$

since F is constant. Thus, if $X \in T_{\pi_1(p)}Z$

$$(\pi^{-1}\nabla^{\alpha})_X F = p[\alpha_p(X'), F_o]p^{-1} \in \text{End}\, T_{\pi_1(p)}N$$

$$= [p\alpha_p(X')p^{-1}, F_{\pi_1(p)}]$$

where $\pi_{1*}X' = X$. But the decomposition

$$o(V) = h^{F_o} + m^{F_o}$$

induces

$$(4.2) \qquad \alpha = \alpha^0 + \alpha^1$$

and so

$$(\pi^{-1}\nabla^{\alpha})_X F = [p\alpha^1_p(X')p^{-1},\ F_{\pi_1(p)}].$$

We shall be done if we prove the

<u>Lemma 4.3</u>

$$P^{\alpha}_{\pi_1 p}(X) = p\alpha^1_p(X')p^{-1} \qquad \text{if} \qquad \pi_{1*}X' = X.$$

Proof By definition $p\alpha^1_p(X')p^{-1}$ vanishes on horizontal vectors, whilst if X is vertical, then $X = pAp^{-1}$ with $A \in m^{F_0}$. Consider $X' = \tilde{A}_p$. Then $\pi_{1*}X' = X$ and

$$\alpha(\tilde{A}) = A$$

gives

$$p\alpha^1_p(X')p^{-1} = pAp^{-1} = X.$$

Note that ad F is invertible on V so Proposition 4.2 determines p^α in terms of $\pi^{-1}\nabla^\alpha F$ and gives a formula for H^α without passing to the principal bundle $O(N,h)$.

Corollary 4.4

$$H^\alpha_F = \{X \in T_FZ : (\pi^{-1}\nabla^\alpha)_X F = 0\}.$$

Of course the horizontal distribution H^α splits the sequence (4.1) and allows us to transfer F to H^α to give F^H. On V we have the endomorphisms J^V determined by the complex structure on the fibre Then on Z we have two natural f-structures $F^\alpha_1 = (F^H, J^V)$, $F^\alpha_2 = (F^H, -J^V)$

Since we have a Riemannian manifold (N,h) we may of course take α to be the Levi-Civita connection λ, in which case we omit the index α and refer to F_1 and F_2. We shall mostly be interested in F_2. For this f-structure

$$T^+Z = H^+ + V'', \quad T^0Z = H^0, \; T^-Z = H^- + V'.$$

If we identify H with E via $d\pi$ we have

$$T^+Z = E^+ + V'', \quad T^0Z = E^0, \quad T^-Z = E^- + V'.$$

Note that since E and V are associated to $\pi_1 \; O(N,h) \longrightarrow Z$, then the connection α makes TZ associated to this bundle with fibre

$V + m^{F_o}$. T^+Z is associated to $O(N,h)$ with fibre $V^+ + m^{-,F_o}$. P^α is an End E-valued 1-form so we can define a new connection D^α in E by

$$D^\alpha = \pi^{-1}\nabla^\alpha - P^\alpha .$$

Proposition 4.5

$$D^\alpha F = 0.$$

Proof If $D^\alpha = \pi^{-1}\nabla^\alpha - P^\alpha$ on E then

$$D^\alpha F = \pi^{-1}\nabla^\alpha F - [P^\alpha, F] = 0$$

by Proposition 4.2

Remark 4.6 D^α is the connection in $E = O(N,h)\times_{H^{F_o}}V$ induced by α^0, the component of α in h^{F_o}, which is a connection in $\pi_1 \quad O(N,h) \to Z$.

When V has even dimension and $2k = \dim V$, then elements of $F_k(V,h_o)$ are non-singular, so satisfy $F^2 = -1$, and are complex structures on V. In this case we denote $F_k(V,h_o)$ by $J(V,h_o)$. Analogously, if N is even dimensional of dimension 2k then we denote $F_k(N,h)$ by $J(N,h)$. $J(N,h)$ is the fibre bundle over N whose sections are almost Hermitian structures on N compatible with the given metric. The canonical section of End $\pi^{-1}TN$ on $J(N,h)$ we denote by J, and the f-structures F_1, F_2 become almost complex structures denoted by J_1, J_2 respectively.

For later use we compute the torsion of D^α. Recall that the torsion T^{D^α} of a connection D on the tangent bundle of a manifold is defined by

$$T^D(X.Y) = D_XY - D_YX - [X,Y]$$

for vector fields X and Y. In particular we have the torsion T^α of ∇^α and T^{D^α} of D^α. By definition the torsion is a tangent bundle valued 2-form, so when the tangent bundle decomposes we can consider the components of the values of the torsion. Thus of $F_k(N,h)$ we have the horizontal and vertical parts of T^{D^α}.
These are determined by $d\pi \circ T^{D^\alpha}$ and $P \circ T^{D^\alpha}$ respectively.

Proposition 4.7

$$d\pi \circ T^{D^\alpha} = \pi^* T^\alpha - P^\alpha \wedge d\pi$$

Proof It suffices to check this identity on vector fields X, Y on $F_k(N,h)$ which are π-related to vector fields X, Y on N Then

$$\begin{aligned} d\pi(T^{D^\alpha}(X,Y)) &= d\pi(D^\alpha_X Y) - d\pi(D^\alpha_Y X) - d\pi[X,Y] \\ &= D^\alpha_X(\pi^{-1}Y) - D^\alpha_Y(\pi^{-1}X) - \pi^{-1}[X,Y] \\ &= (\pi^{-1}\nabla^\alpha)_X \pi^{-1}Y - (\pi^{-1}\nabla^\alpha)_Y \pi^{-1}X - \pi^{-1}[X,Y] \\ &\quad - P^\alpha(X)d\pi Y + P^\alpha(Y)d\pi X \\ &= \pi^* T^\alpha(X,Y) - P^\alpha(X)d\pi Y + P^\alpha(Y)d\pi X \end{aligned}$$

which proves the Proposition.

The vertical part can be found by using Proposition 4.2 and 4.5:

$$\begin{aligned} &[P^\alpha(T^{D^\alpha}(X,Y)), F] \\ &\quad = [D^\alpha_X P^\alpha(Y) - D^\alpha_Y P^\alpha(X) - P^\alpha([X,Y]), F] \\ &\quad = D^\alpha_X[P(Y),F] - D^\alpha_Y[P^\alpha(X),F] - [P^\alpha([X,Y]),F] \\ &\quad = (\pi^{-1}\nabla^\alpha)_X(\pi^{-1}\nabla^\alpha)_Y - [P^\alpha(X),[P^\alpha(Y),F]] \end{aligned}$$

$$- (\pi^{-1}\nabla^{\alpha})_Y(\pi^{-1}\nabla^{\alpha})_X + [P^{\alpha}(Y),[P^{\alpha}(X), F]]$$

$$- (\pi^{-1}\nabla^{\alpha})_{[X,Y]} F$$

$$= [\pi^*R^{\alpha}(X,Y), F] - [[P^{\alpha}(X),P^{\alpha}(Y)]\ F].$$

If we view $P^{\alpha} \circ T^{D^{\alpha}}$ as an $\mathrm{End}\,\pi^{-1}TN$-valued 2-form then it equals the component of $\pi^*R^{\alpha} - \frac{1}{2}[P^{\alpha} \wedge P^{\alpha}]$ in V. Given an element A of $\mathrm{End}\,\pi^{-1}TN$ we denote its component in V by A^m. Then we have proven

<u>Proposition 4.8</u>

$$P^{\alpha} \circ T^{D^{\alpha}} = \{\pi^*R^{\alpha} - \tfrac{1}{2}[P^{\alpha} \wedge P^{\alpha}]\}^m.$$

5. MAPS INTO $F_k(N,h)$

In this section we shall study what it means for maps of complex manifolds into $F_k(N,h)$ to be holomorphic. If π is the projection to N and ψ a smooth map of some manifold M into $F_k(N,h)$ we denote by φ the composite map $\pi \circ \psi$ from M to N. Then $\psi(x) \in \pi^{-1}(x)$ so is an f-structure of rank 2k on $T_{\varphi(x)}N$. Thus ψ can be viewed as an f-structure on $\varphi^{-1}TN$, and when thought of as such we shall denote it by F^{ψ}. Since $\varphi^{-1}TN = \psi^{-1} \circ \pi^{-1}TN = \psi^{-1}E$ it is clear that $F^{\psi} = \psi^{-1}F$. Alternatively we may think of F^{ψ} in terms of its $+i$ eigenbundle in $\varphi^{-1}TN^{\mathbb{C}}$ which we denote by ψ^{+}. ψ^{+} is a rank k complex subbundle of $\varphi^{-1}TN^{\mathbb{C}}$ which is isotropic with respect to the bilinear extension of $\varphi^{-1}h$. Clearly $\psi^{+} = \psi^{-1}E^{+}$ and ψ is determined by φ and $\psi^{+} \subset \varphi^{-1}TN^{\mathbb{C}}$. We shall translate conditions on ψ into conditions on φ and F^{ψ} or ψ^{+}.

Our first task is to determine $d\psi$ which is a 1-form on M with values in $\psi^{-1}TF_k(N,h)$. If we fix a metric connection α on TN giving

$$P^{\alpha} : TZ \longrightarrow V$$

and an isomorphism

$$TZ \cong V + E;$$

then we get a corresponding decomposition of $d\psi$ into V-valued and E-valued 1-forms. By the chain rule $d\pi \circ d\psi = d\varphi$, so the part in $\psi^{-1}E = \psi^{-1} \circ \psi^{+1}TN = \varphi^{-1}TN$ is just $d\varphi$ whilst the part in V can obviously be identified with $\psi^{*}P^{\alpha}$ viewing P^{α} as a V-valued 1-form. Thus we have

Proposition 5.1 $d\psi = (\psi^*P^\alpha, d\varphi)$ relative to the decomposition $TF_k(N,h) \cong V + E$.

ψ^*P^α is easily determined from Proposition 4.1. Viewing V as a subbundle of End E we have

$$[P^\alpha, F] = (\pi^{-1}\nabla^\alpha)F$$

so

$$[\psi^*P^\alpha, \psi^{-1}F] = (\varphi^{-1}\nabla^\alpha)\psi^{-1}F,$$

and since $\psi^{-1}F = F^\psi$ we have shown:

Proposition 5.2 $[(\psi^*P^\alpha)(X), F^\psi] = (\varphi^{-1}\nabla^\alpha)_X F^\psi, X \in TM.$

Since bracketing with F^ψ is injective on $V \subset$ End E, it follows $d\psi(X)$ is horizontal if and only if $(\varphi^{-1}\nabla^\alpha)_X F^\psi = 0$. By Lemma 2.3 this holds if and only if $(\varphi^{-1}\nabla^\alpha)_X C^\infty(\psi^+) \subset C^\infty(\psi^+)$. Thus we have shown:

Corollary 5.3 The following conditions on a map $\psi : M \to F_k(N,h)$ are equivalent:

(i) ψ is horizontal for α (that is $d\psi(T_xM) \subset H^\alpha_{\psi(x)}$, $\forall x \in M$);

(ii) $(\varphi^{-1}\nabla^\alpha)F^\psi = 0$;

(iii) $(\varphi^{-1}\nabla^\alpha)_X C^\infty(\psi^+) \subset C^\infty(\psi^+)$, for all vectorfields X on M.

We can apply the same methods to study when a map ψ of an almost complex manifold M into $F_k(N,h)$ is f-holomorphic with respect to F_1^α or F_2^α.

Theorem 5.4 The following conditions on a map ψ of the almost complex manifold M into $F_k(N,h)$ are equivalent:

(i) ψ is F_1^α-holomorphic;

(ii) (a) $d\varphi \circ J^M = F^\psi \circ d\varphi$ and

(b) $(\varphi^{-1}\nabla^\alpha)_X(F^\psi)\psi^+ = 0$ for X in $T'M$;

(iii) (a) $d\varphi(T'M) \subset \psi^+$ and

(b) $(\varphi^{-1}\nabla^\alpha)_X C^\infty(\psi^+) \subset C^\infty(\psi^+)$ for X in $T'M$.

<u>Theorem 5.5</u> The following conditions on a map ψ of the almost complex manifold M into $F_k(N,h)$ are equivalent:

(i) ψ is F_2^α-holomorphic;

(ii) (a) $d\varphi \circ J^M = F^\psi \circ d\varphi$ and

(b) $(\varphi^{-1}\nabla^\alpha)_X(F^\psi)\psi^+ = 0$ for X in $T''M$;

(iii) (a) $d\varphi(T'M) \subset \psi^+$ and

(b) $(\varphi^{-1}\nabla^\alpha)_X C^\infty(\psi^+) \subset C^\infty(\psi^+)$ for X in $T''M$.

We shall prove Theorem 5.5, the proof of Theorem 5.4 being similar.

<u>Proof of Theorem 5.5</u> By definition, ψ is F_2^α-holomorphic if and only if

$$d\psi \circ J^M = \psi^{-1}F_2^\alpha \circ d\psi$$

Splitting into horizontal and vertical parts we get

$$d\varphi \circ J^M = \psi^{-1} F \circ d\varphi$$

which is (iia) and

$$\psi^*P^\alpha \circ J^M = -\psi^{-1}J^V \circ \psi^*P^\alpha .$$

If we evaluate both sides on a vector X and bracket with F^ψ we get

$$(\varphi^{-1}\nabla^\alpha)_{J^M X}F^\psi = -\psi^{-1}J^V(\phi^{-1}\nabla^\alpha)_X F^\psi .$$

For X in $T''M$ this will be so if and only if

$$\psi^{-1}J^{V}(\varphi^{-1}\nabla^{\alpha})_X F^{\psi} = i(\psi^{-1}\nabla^{\alpha})_X F^{\psi}$$

or

$$(\varphi^{-1}\nabla^{\alpha})_X F^{\psi} \in \psi^{-1}V'.$$

By Proposition 3.3, if we put $A = (\varphi^{-1}\nabla^{\alpha})_X F^{\psi}$, this is equivalent to

$$A\psi^{+} = 0, \quad A\psi^{0} \subset \psi^{+}, \quad A\psi^{-} \subset \psi^{0} + \psi^{+}$$

where $\psi = \psi^{-1}E$, $\psi^{0} = \psi^{-1}E^{0}$. But for A of this form the last condition is automatic by Proposition 2.5, and the second follows from the first which is (iib). Thus we have shown that (i) and (ii) are equivalent. (iii) is a straightforward translation of (ii) in terms of bundles

Condition (iiib) strongly resembles condition A if we take α to be Levi-Civita connection, so it is not surprising that we have

<u>Theorem 5.6</u> If (M,g,J^{M}) is almost Hermitian with coclosed Kaehler form and $\psi : M \longrightarrow F_k(N,h)$ is F_2-holomorphic then $\varphi = \pi \circ \psi$ is harmonic.

<u>Proof</u> From Theorem 5.5 we know ψ F_2-holomorphic implies

$$d\psi(T'M) \subset \psi^{+}, \quad (\varphi^{-1}\nabla)_X C^{\infty}(\psi^{+}) \subset C^{\infty}(\psi^{+})$$

for X in $T''M$. We proceed exactly as in the proof of Theorem 2.6 and show

$$\tau_{\varphi} = \sum_j \nabla d\varphi(\overline{Z}_j, Z_j) \in C^{\infty}(\psi^{+}).$$

But τ_{φ} is real, so has values in $\psi^{+} \cap \overline{\psi^{+}} = \psi^{+} \cap \psi^{-} = 0$. Thus $\tau_{\varphi} = 0$ and so φ is harmonic as claimed.

Remark 5.7 If we further assume M is (1,2)-symplectic then as in section 2 we can conclude that $(\nabla d\varphi)^{1,1}$ vanishes.

When N has even dimension $2k$ then we denoted $F_k(N,h)$ by $J(N,h)$ in the previous section to indicate we are dealing with genuine complex structures, and a map $\psi : M \longrightarrow J(N,h)$ corresponds with a complex structure J^{ψ} on $\varphi^{-1}TN$ or a maximal isotropic subbundle $\psi^{+} \subset \varphi^{-1}TN^{\mathbb{C}}$. We summarize the results of this section in this particular case in

Theorem 5.7 If $\psi : M \longrightarrow J(N,h)$ is a smooth map of the almost complex manifold (M,J^M) and $\varphi = \pi \circ \psi$, $J^{\psi} = \psi^{-1}g$ then

(i) ψ is horizontal $\iff (\varphi^{-1}\nabla)J^{\psi} = 0 \iff (\varphi^{-1}\nabla)C^{\infty}(\psi^{+}) \subset C^{\infty}(\psi^{+})$;

(ii) ψ is J_1-holomorphic $\iff d\varphi \circ J^M = J^{\psi} \circ d\varphi$ and $(\varphi^{-1}\nabla)_X C^{\infty}(\psi^{+}) \subset C^{\infty}(\psi^{+})$ for X in $T'M$;

(iii) ψ is J_2-holomorphic $\iff d\varphi \circ J^M = J^{\psi} \circ d\varphi$ and $(\varphi^{-1}\nabla)_X C^{\infty}(\psi^{+}) \subset C^{\infty}(\psi^{+})$ for X in $T''M$.

If M is almost Hermitian with coclosed Kaehler form and ψ is J_2-holomorphic then φ is harmonic. If M is Kaehler then $(\nabla d\varphi)^{1,1}$ vanishes.

It is natural to ask if we can start with φ and find a lift ψ into some $F_k(N,h)$? If φ is an immersion and M is almost Hermitian of dimension $2k$, then $d\varphi(T'_xM)$ will be a k-dimensional subspace of $T_{\varphi(x)}N^{\mathbb{C}}$ but may not be isotropic. The condition of isotropy is:

$$h_{\varphi(x)}(d\varphi(T'_xM), d\varphi(T'_xM)) = 0$$

or $(\varphi^*h)^{2,0}_x = 0$. Thus if $(\varphi^*h)^{2,0} = 0$ then $d\varphi(T'M)$ is an isotropic k-dimensional subbundle of $\varphi^{-1}TN^{\mathbb{C}}$ and so yields a map $\tilde{\varphi} : M \to F_k(N,h)$ Since $\tilde{\varphi}^{+} = d\varphi(T'M)$, condition (iiia) of Theorem 5.5 is automatically satisfied, so $\tilde{\varphi}$ is F_2-holomorphic if and only if (iiib) holds for $\tilde{\varphi}^{+}$. This last condition is equivalent to

$$(\varphi^{-1}\nabla)_X\, d\varphi(Y) \in d\varphi(T'M)$$

for X in $T''M$, Y in $T'M$. If M is (1,2)-symplectic we see this is the same as

$$(\nabla d\varphi)(X,Y) \in d\varphi(T'M), \quad X \in T''M, \quad Y \in T'M;$$

so by the usual argument this is equivalent to $(\nabla d\varphi)^{1,1} = 0$. Thus we have

Theorem 5.8 If (M^{2k},g,J) is (1,2)-symplectic and $\varphi : M \longrightarrow N$ an immersion with $(\varphi^* h)^{2,0} = 0$ then $\tilde{\varphi} : M \to F_k(N,h)$ defined by $\tilde{\varphi}^+ = d\varphi(T'M)$ is F_2-holomorphic if and only if $(\nabla d\varphi)^{1,1} = 0$.

Remark 5.9 If ψ is any other lift to $F_k(N,h)$ which is F_2-holomorphic then (iiia) of Theorem 5.5 shows $\tilde{\varphi}^+ \subset \psi^+$ and since they are of the same rank $\tilde{\varphi}^+ = \psi^+$ so $\tilde{\varphi} = \psi$ and hence the lift is unique with the holomorphic property.

Remark 5.10 For $k = 1$ the restriction that φ is an immersion can be removed. This case was considered by Eells-Salamon [4,11] and is described in the next section.

6. CONFORMAL HARMONIC MAPS OF RIEMANN SURFACES

Let M be a Riemann surface. That is, M is an oriented surface with a conformal class of metrics. The conformal class is fixed by specifying a complex structure J on M (there is no obstruction to integrability in two dimensions) relative to which the metrics are Hermitian Since any 3-form on M vanishes identically, each metric in this conformal class is Kaehler. Moreover the harmonic map equation

$$\tau_\varphi = 0$$

is conformally invariant and relative to a holomorphic coordinate z on M takes the simple form

$$(\varphi^{-1}\nabla)_{\partial/\partial\bar{z}}\ \varphi_*\frac{\partial}{\partial z} = 0 \tag{6.1}$$

It is extremely useful to interpret this equation as saying $\varphi_*\ \partial/\partial z$ is a holomorphic section of $\varphi^{-1}TN^{\mathbb{C}}$. For this we need the following result of Koszul and Malgrange [6].

Theorem 6.1 Let E be a complex vector bundle over a Riemann surface M and D a linear connection in E then E has a unique holomorphic structure such that a section s of E over an open set U is holomorphic if and only if $D_X s = 0$ for every (0,1) vectorfield X on U.

Remark 6.2 Note that for the holomorphic structure arising from this theorem the $\bar{\partial}$-operator is the (0,1) part D" of the connection D. Note also that a smooth subbundle F of E is a holomorphic subbundle of E if and only if $D''C^\infty(F) \subset C^\infty(F)$.

Remark 6.3 Theorem 5.5 now translates as: a map ψ of a Riemann surface M into $F_k(N,h)$ is F_2-holomorphic if and only if $d\varphi(T'M) \subset \psi^+$ and ψ^+ is a holomorphic subbundle of $\varphi^{-1}TN^{\mathbb{C}}$ for the Koszul-Malgrange holomorphic structure.

Remark 6.4 We also need the following observation. if E is a holomorphic vector bundle and F a holomorphic line bundle over a Riemann surface and s a non-zero holomorphic section of $F \otimes E$ then there is a unique homorphic line subbundle $L \subset E$ such that s is a section of $F \otimes L$. This is shown as follows: where $s(x) = 0$, if s_o is a local holomorphic section of F non-zero at x then $s = s_o \otimes t$ near x for a local holomorphic section t of E which is also non-zero at x. Then $L_x = \mathbb{C}t(x)$ is independent of the choice of s_o and t. The zeros of s are isolated, and of finite order, thus near any such zero we have $s = z^k s'$ with s' zero-free and holomorphic. Then s' determines the fibre of L at the zero just as s does at other points. Since we have given L locally by zero-free holomorphic sections, L is clearly a holomorphic line subbundle.

Now suppose

$$\varphi \cdot M \longrightarrow N$$

is a harmonic map of the Riemann surface M into the Riemannian manifold (N,h). The differential $d\varphi$ is a 1-form with values in $\varphi^{-1}TN$, and its $(1,0)$ part $\partial\varphi$ is a 1-form with values in $\varphi^{-1}TN^{\mathbb{C}}$. $\varphi^{-1}TN$ and its complexification have the pull-back connection $\varphi^{-1}\nabla$ to which we can apply Theorem 6.1 to give $\varphi^{-1}TN^{\mathbb{C}}$ a holomorphic structure with $(\varphi^{-1}\nabla)''$ as the $\bar{\partial}$-operator. Then (6.1) says φ harmonic implies $\partial\varphi$ is a holomorphic section of $K_M \otimes \varphi^{-1}TN^{\mathbb{C}}$ which is non-zero if φ is non constant. By Remark 6.3 this gives a holomorphic line bundle $\tilde{\varphi}^+ \subset \varphi^{-1}TN^{\mathbb{C}}$. φ is conformal if φ^*h is type $(1,1)$, or $\varphi(T'M)$ is an isotropic line in $\varphi^{-1}TN^{\mathbb{C}}$ for the complex bilinear extension of $\varphi^{-1}h$. This means $\tilde{\varphi}^+$ is an isotropic line in $\varphi^{-1}TN^{\mathbb{C}}$. Thus we have

Theorem 6.5 A non-constant harmonic map $\varphi : M \longrightarrow N$ gives a holomorphic line subbundle $\tilde{\varphi}^+ \subset \varphi^{-1}TN^{\mathbb{C}}$ which is spanned by $d\varphi(T'M)$ where this is non-zero. $\tilde{\varphi}^+$ is isotropic if and only if φ is conformal.

If φ is conformal and harmonic then we get a map

$$\tilde{\varphi} : M \to F_1(N,h)$$

with $\varphi = \pi \circ \tilde{\varphi}$. Since the conditions of Theorem 5.5 are satisfied $\tilde{\varphi}$ is F_2-holomorphic. Conversely, suppose that $\psi : M \to F_1(N,h)$ is F_2-holomorphic, then we have $\tilde{\varphi}^+ \subset \psi^+$ and ψ^+ is holomorphic. If φ is non-constant so ψ is non-vertical, then $\tilde{\varphi}^+$ and ψ^+ are both line bundles so $\tilde{\varphi}^+ = \psi^+$. φ is then both conformal and harmonic. Thus we have the following result of Eells and Salamon.

Theorem 6.6 There is one-one correspondence between non-constant (weakly) conformal harmonic maps φ $M \longrightarrow N$ and non-vertical F_2-holomorphic maps $\psi : M \longrightarrow F_1(N,h)$. The correspondence is determined by $\varphi = \pi \circ \psi$ and $\psi = \tilde{\varphi}$.

If we want to get analogous results using other spaces such as $F_k(N,h)$ we need to build lifts into these spaces, that is to build isotropic subbundles of $\varphi^{-1}TN^{\mathbb{C}}$. By comparison with the work of Eells and Wood [5] we do this by taking successive derivatives of $\partial\varphi$. This introduces a notion of isotropy for maps of a Riemann surface which generalizes the notion of being conformal. In addition, in order to extend over singularities as we did in Remark 6.4, we introduce a notion we call curvature isotropy which is automatic for harmonic maps into a space of constant curvature. We do not know of any non-trivial example of a curvature isotropic map except in the constant curvature case. The results of this part were also obtained independently by F. Burstall

Thus let $\varphi . M \longrightarrow N$ be a smooth map of the Riemann surface M, and for a holomorphic coordinate z on M let

$$\partial\varphi = \delta_z\varphi \otimes dz, \quad \bar{\delta}_z\varphi = \overline{\delta\varphi},$$

and

$$\delta = (\varphi^{-1}\nabla)_{\partial/\partial z}, \quad \bar{\delta} = (\varphi^{-1}\nabla)_{\partial/\partial \bar{z}}.$$

Put

$$(W^{\alpha}_{\varphi})_z = \mathrm{Span}_{\mathbb{C}}\{\delta\varphi_z, \ldots, \delta^{\alpha}\varphi_z\}, \quad \alpha = 1,2,\ldots$$

and

$$(\bar{W}^{\alpha}_{\varphi})_z = \overline{(W^{\alpha}_{\varphi})_z}.$$

The order of φ is defined to be the maximum dimension (over z and α) of the spaces $(W^{\alpha}_{\varphi})_z$. If φ has order k then $(W^{k}_{\varphi})_z$ has dimension k for z in a non-empty open set. It is clear that each W^{α}_{φ} is independent of the coordinates used to define it for each α.

We say φ is α-real isotropic if W^{α}_{φ} is a family of isotropic subspaces of $\varphi^{-1}TN^{\mathbb{C}}$ with respect to $\varphi^{-1}h$. Since $W^{1}_{\varphi} = d\varphi(T'M)$, this generalizes the notion of being conformal φ is real isotropic if α-real isotropic for all α. This is the same as k-real isotropic if k is the order of φ.

If we put

$$R_o = \varphi^{*}R^{N}(\partial/_{\partial z}, \partial/_{\partial \bar{z}})$$

we say φ is curvature isotropic if

$$R_o W^{\alpha}_{\varphi} \subset W^{\alpha}_{\varphi}, \quad \alpha = 1,2,\ldots$$

This is also independent of coordinates, and it can be shown that if φ is harmonic and N of constant sectional curvature then φ is curvatu re isotropic

If φ is harmonic, real isotropic and curvature isotropic then we can fill out W^k_φ to an isotropic subbundle of $\varphi^{-1}TN^{\mathbb{C}}$ just as we extended $\partial\varphi$ to $\tilde{\varphi}$. This is because if we consider

$$\delta\varphi \wedge \delta^2\varphi \wedge \ldots \wedge \delta^k\varphi$$

as a section of $\Lambda^k\varphi^{-1}TN^{\mathbb{C}}$ then

$$\bar{\delta}(\delta\varphi \wedge \ldots \wedge \delta^k\varphi) = \bar{\delta}\delta\varphi \wedge \delta^2\varphi \wedge \ldots \delta^k\varphi + \ldots + \delta\varphi \wedge \ldots \wedge \bar{\delta}\delta^k\varphi.$$

Now

$$\bar{\delta}\delta^\alpha\varphi = [\bar{\delta},\delta]\delta^{\alpha-1} + \delta[\bar{\delta},\delta]\delta^{\alpha-1}\varphi + \ldots + \delta^{\alpha-1}\bar{\delta}\delta\varphi \in W_\varphi^{\alpha-1}$$

since $[\bar{\delta},\delta] = -R_o$ and φ is curvature isotropic Thus

$$\bar{\delta}(\delta\varphi \wedge \ldots \wedge \delta^k\varphi) = 0$$

and so $\delta\varphi \wedge \ldots \wedge \delta^k\varphi$ is a holomorphic section of $\Lambda^k\varphi^{-1}TN^{\mathbb{C}}$. It the determines, by Remark 6 4, a holomorphic line subbundle of $\Lambda^k\varphi^{-1}TN^{\mathbb{C}}$ which is decomposible on a dense open set and hence everywhere. This gives a rank k subbundle of $\varphi^{-1}TN^{\mathbb{C}}$, isotropic on a dense open set, and hence everywhere In fact, by construction, this is a holomorphic isotropic subbundle $\tilde{\varphi}^+_k \subset \varphi^{-1}TN^{\mathbb{C}}$ Further, since $\delta\varphi \in \varphi^+_k$, we have $\tilde{\varphi}^+ \subset \tilde{\varphi}^+_k$ and hence as a map

$$\tilde{\varphi}_k : M \longrightarrow F_k(N,h)$$

$\tilde{\varphi}_k$ is F_2-holomorphic. Thus we have shown.

<u>Theorem 6.7</u> A harmonic, real isotropic, curvature isotropic map $\tilde{\varphi} : M \longrightarrow N$ determines an F_2-holomorphic map

$$\tilde{\varphi}_k . M \longrightarrow F_k(N,h)$$

where k is the order of φ. $\tilde{\varphi}_k$ agrees with W^k_φ where the latter has maximal rank.

In fact we can show that $\tilde{\varphi}_k$ satisfies the stronger condition of being horizontal as a map. This is because

$$\bar{\delta}\, C^\infty(\tilde{\varphi}_k^+) \subset C^\infty(\tilde{\varphi}_k^+)$$

since φ_k^+ is holomorphic as a bundle and

$$\delta\, C^\infty(\tilde{\varphi}_k^+) \subset C^\infty(\tilde{\varphi}_k^+)$$

by construction. Thus

$$\varphi^{-1}\nabla_X\; C^\infty(\tilde{\varphi}_k^+) \subset C^\infty(\tilde{\varphi}_k^+)$$

for all vector fields X on M. Thus we have

Theorem 6.8 The lift $\tilde{\varphi}_k$ in Theorem 6.7 is horizontal.

A partial converse is also true.

Theorem 6.9 If $\psi : M \to F_k(N,h)$ is horizontal and F_2-holomorphic then $\varphi = \pi \circ \psi$ is real isotropic, harmonic and has order at most k.

Proof Theorem 5.5 tells us that φ is harmonic. ψ is F_2-holomorphic implies $\tilde{\varphi}^+ \subset \psi^+$ and ψ horizontal implies $\delta\, C^\infty(\psi^+) \subset C^\infty(\psi^+)$. Then $\delta\,\varphi \in \psi^+$ so $\delta^\alpha \varphi \in \psi^+$ for all α, so $W^\alpha_\varphi \subset \psi^+$ for all α. Since ψ^+ is isotropic, so is W^α_φ and hence φ is real isotropic. Since ψ^+ has rank k, $\dim W^\alpha_\varphi \leqslant k$ for all α so order $\varphi \leqslant k$.

In order to get a one-one correspondence as we had in Theorem 6.6 we need a notion of curvature isotropic for map $\psi . M \to F_k(N,h)$. This can be acheived using the connection D, which is induced from $\pi^{-1}\nabla - P$ on E. Thus we can form

$$\delta\psi = d\psi(\partial/_{\partial z})$$

and

$$\delta = (\psi^{-1}D)_{\partial/_{\partial z}} \quad , \quad \bar{\delta} = (\psi^{-1}D)_{\partial/_{\partial \bar{z}}} \quad .$$

This gives us

$$(W^{\alpha}_{\varphi})_z = \mathrm{Span}_{\mathbb{C}}\{\delta\psi, \delta^2\psi, \ldots, \delta^{\alpha}\psi\}, \quad \alpha = 1,2,\ldots$$

and we say ψ has order k if this is the maximum dimension of all the $(W^{\alpha}_{\psi})_z$. We say ψ is curvature isotropic, if

$$\psi^* R^D(\partial/_{\partial z}, \partial/_{\partial \bar{z}})(W^{\alpha}_{\psi})_z \subset (W^{\alpha}_{\psi})_z \ , \quad \forall z,\alpha$$

where R^D is the curvature of D.

Theorem 6.10 There is a one-one correspondence between curvature isotropic horizontal F_2-holomorphic maps $\psi : M \to F_k(N,h)$ of order k and harmonic, real isotropic, curvature isotropic maps $\tilde{\varphi} : M \to N$ of order k given by $\varphi = \pi \circ \psi$, $\psi = \varphi_k$.

Proof If $\psi : M \to F_k(N,h)$ is horizontal, F_2-holomorphic, then $W^{\alpha}_{\varphi} \subset \psi^+$ for all α. Since ψ is horizontal, and $H \cong \pi^{-1}TN$ then $\delta\psi = \delta\varphi$ under this isomorphism and $\psi^{-1}D = \varphi^{-1}\nabla$, so

$$d\pi\ \delta^{\alpha}\psi = \delta^{\alpha}\varphi \quad , \quad \alpha = 1,2,\ldots$$

and $d\pi$ maps W^{α}_{ψ} isomorphically to W^{α}_{φ}. Thus W^{α}_{φ} is preserved by R_o since

$$R_o \circ d\pi = d\pi \circ \psi^* R^D(\partial/_{\partial z}, \partial/_{\partial \bar{z}}).$$

Then φ is order k and curvature isotropic. This gives $\tilde{\varphi}_k$ as the extension of W^k_{φ} and $\tilde{\varphi}^+_k \subset \psi^+$, so $\tilde{\varphi}^+_k = \psi^+$ since both have rank k. The converse is given by Theorems 6.7 and 6.8.

7. GENERALIZED F-TWISTOR SPACES

The f-structures F_1 and F_2 we defined on $F_k(N,h)$ can be defined on other fibre bundles with similar properties. In fact we need very little structure to be able to make the definitions work, we shall however assume rather more in order to be able to make calculations. Thus suppose Z is a smooth fibre bundle over N with each fibre a complex manifold whose complex structure varies smoothly from fibre to fibre. In addition we assume we have a fibre preserving map

$$Z \xrightarrow{f} F_k(N,h), \tag{7.1}$$

such that f is holomorphic on each fibre, and Z has a horizontal distribution. We can then define J^V and F^H, and set

$$F_1 = (J^V, F^H), \quad F_2 = (-J^V, F^H)$$

just as we did for $F_k(N,h)$. If f maps the horizontal distribution on Z to that of the Levi-Civita connection on $F_k(N,h)$ we say Z is a metric f-twistor space. Then f is a morphism of f-structures for both F_1 and F_2. Any F_2-holomorphic map of a complex manifold into Z yields, by composition with f, an F_2-holomorphic map into $F_k(N,h)$. If the domain has coclosed Kaehler 2-form we shall then obtain a harmonic map into N.

Examples

(i) A general class of examples can be obtained as follows: suppose the tangent bundle TN of N is associated to a principal G-bundle P by an orthogonal representation V of G,

$$TN = P \times_G V.$$

Let C be any conjugacy class in the Lie algebra $\mathfrak{g}$ of G and

$$Z = P \times_G C.$$

Each ξ in C determines a skew-symmetric endomorphism of V, so decomposes $V^{\mathbb{C}}$ into eigenspaces

$$V^{\lambda,\xi} = \{v \in V^{\mathbb{C}} \quad \xi v = i\lambda v\}.$$

We put

$$V^{+,\xi} = \sum_{\lambda>0} V^{\lambda,\xi}, \quad V^{-,\xi} = \sum_{\lambda<0} V^{\lambda,\xi} = \overline{V^{+,\xi}}$$

so

$$V^{\mathbb{C}} = V^{+,\xi} + V^{0,\xi} + V^{-,\xi}$$

is a decomposition of $V^{\mathbb{C}}$ stable under the stability group of ξ in C. Moreover, if

$$\mathfrak{g}^{\lambda,\xi} = \{\eta \in \mathfrak{g}^{\mathbb{C}} : [\xi,\eta] = i\lambda\eta\}$$

then

$$\mathfrak{g}^{\lambda,\xi} V^{\mu,\xi} \subset V^{\lambda+\mu,\xi}$$

so the subspace $\mathfrak{m}^{+,\xi}$ which defines the complex structure on C (see section 3) will satisfy

$$\mathfrak{m}^{+,\xi} V^{+,\xi} \subset V^{+,\xi}.$$

Let $k = \dim V^{+,\xi}$. We have a holomorphic map

$$f_o : C \longrightarrow F_k(V,h_o)$$

given by

$$f_o(\xi) = \begin{cases} 1 \text{ on } V^{+,\xi} \\ 0 \text{ on } V^{0,\xi} \\ -1 \text{ on } V^{-,\xi} \end{cases}$$

which is G-equivariant, and so induces a map

$$f : Z \longrightarrow F_k(N,h)$$

It is not hard to verify that any connection α in P induces compatible horizontal distributions on Z and $F_k(N,h)$. If α induces the Levi-Civita connection on TN we will thus have a metric f-twistor space.

(ii) A special case of (i) would be to take N to be a Kaehler manifold of complex dimension n with P the unitary frame bundle and C the Grassmannian of k-planes in $\mathbb{C}^n$. This Grasmannian can be viewed as a conjugacy class in $u(n)$ in more than one way each of which gives $G_k(T'N)$ an interpretation as an f-twistor space. A k-plane W in $T'N$ can be viewed also as a J-stable real 2k-plane in TN, and as such we can extend J outside W by 0. This gives the first interpretation and gives a map

$$f : G_k(T'N) \longrightarrow F_k(N,h)$$

The second interpretation corresponds with extending J outside W as $-J$; that is we set

(7.2)
$$j(W) = \begin{cases} J \text{ on } W \\ -J \text{ on } W^{\perp} \end{cases}$$

to give the map

$$j : G_k(T'N) \longrightarrow J(N,h).$$

(iii) Another special case of (i) occours if N is a homogeneous Riemannian manifold K/G, then for H the centralizer in G of an element of the Lie algebra of G the coset fibring

$$K/H \longrightarrow K/G$$

is an f-twistor space. This gives many examples of f-twistor spaces for instance over symmetric spaces which can be studied by Lie-theoretic methods. If H is also the centralizer in K then the f-structure on K/H is almost complex. Such examples have been studied by Salamon [11] and Bryant [2].

8 ISOTROPIC HARMONIC MAPS INTO KAEHLER MANIFOLDS

In section 6 we saw that in order to construct a lift into some $F_k(N,h)$ with $k \geq 1$, we had to assume notions of real and curvature isotropy. In this section we investigate a notion of isotropy appropriate for constructing a lift into $J(N,h)$ when N is a Kaehler manifold (we assume the latter throughout this section). This uses the twistor space structure given by

$$\jmath : G_k(T'N) \longrightarrow J(N,h)$$

defined in the previous section. If

$$\psi : M \longrightarrow G_k(T'N)$$

we denote the composition $\jmath \circ \psi$ by Ψ. Since $\jmath$ is easily seen to be an embedding, ψ is horizontal or holomorphic if and only if Ψ is. For maps into $J(N,h)$, Theorem 5.7 gives the conditions for Ψ to be horizontal or holomorphic, and we now want to translate these back in terms of ψ.

If ψ is a map into $G_k(T'N)$ and φ the induced map into N, we may view ψ as a subbundle of $\varphi^{-1}T'N$, and $\psi^{\perp}$ will denote its orthogonal complement in $\varphi^{-1}T'N$. Then

$$\varphi^{-1}T'N = \psi + \psi^{\perp} , \quad \varphi^{-1}T''N = \bar{\psi} + \overline{\psi^{\perp}} .$$

Moreover from (7.2) we have

$$\psi^{+} = \psi + \overline{\psi^{\perp}} , \quad \psi^{-} = \bar{\psi} + \psi^{\perp} .$$

Proposition 8.1 ψ is horizontal if and only if the space of sections $C^\infty(\psi)$ is stable under $(\varphi^{-1}\nabla)_X$ for all vector fields X on M.

Proof We know that ψ is horizontal if and only if Ψ is, and Ψ is horizontal if and only if $C^\infty(\Psi^+)$ is $(\varphi^{-1}\nabla)_X$-stable for all vector fields X. H Kaehler means $\varphi^{-1}T'N \subset \varphi^{-1}TN^{\mathbb{C}}$ is $(\varphi^{-1}\nabla)_X$-stable, and

$$\psi = \Psi^+ \subset \varphi^{-1}T'N,$$

and so $C^\infty(\Psi^+)$ $(\varphi^{-1})_X$-stable implies $C^\infty(\psi)$ $(\varphi^{-1}\nabla)_X$-stable. The converse follows since ∇ metric and $C^\infty(\psi)$ $(\varphi^{-1}\nabla)_X$-stable implies $C^\infty(\psi^\perp)$ $(\varphi^{-1}\nabla)_X$-stable.

Similar arguments allow us to translate J_2-holomorphicity of ψ. By Theorem 5.7 we must have Ψ^+ a holomorphic subbundle of $\varphi^{-1}TN^{\mathbb{C}}$ and $\delta\varphi$ a section of Ψ^+. If we write $\nabla'\varphi$ and $\nabla''\varphi$ for the parts of $d\varphi$ from $T'M$ and $T''M$ to $\varphi^{-1}T'N$ respectively then we see we must have

$$\nabla'\varphi(T'M) \subset \psi\,, \quad \nabla''\varphi(T''M) \subset \overline{\psi}^\perp.$$

On the other hand, ψ is a parallel subbundle of Ψ^+, so Ψ^+ holomorphic implies ψ holomorphic and conversely since ψ holomorphic implies $\overline{\psi^\perp}$ holomorphic. Thus

Proposition 8.2 A map ψ $M \to G_k(T'N)$ is J_2-holomorphic if and only if $\psi \subset \varphi^{-1}T'N$ is a holomorphic subbundle, and $\nabla'\varphi(T'M) \subset \psi$, $\nabla''\varphi(T''M) \perp \psi$.

In section 6 isotropy lead to considering maps which were both horizontal and holomorphic (J_1 and J_2 agree on horizontal vectors so for horizontal maps we do not need to specify which almost complex structure we are using). If we do the same for maps to $G_k(T'N)$ we have

Proposition 8.3 A map $\psi : M \longrightarrow G_k(T'N)$ is horizontal and holomorphic if and only if $C^\infty(\psi)$ is $(\varphi^{-1}\nabla)_X$-stable for all vector fields X on M, and

$$\nabla'\varphi(T'M) \subset \psi \ , \quad \nabla''\varphi(T''M) \subset \psi \ ,$$

where $\varphi = \pi \circ \psi$.

Suppose now that M is a Riemann surface, then for a holomorphic coordinate z on M we put

$$\nabla'_z = (\varphi^{-1}\nabla)_{\partial/\partial z} \ , \quad \nabla''_{\bar z} = (\varphi^{-1}\nabla)_{\partial/\partial \bar z} \ ,$$

$$\nabla'_z\varphi = \nabla'\varphi(\partial/\partial z) \ , \quad \nabla''_{\bar z}\varphi = \nabla''\varphi(\partial/\partial \bar z),$$

and we define recursively

$$(\nabla'_z)^{k+1}\varphi = \nabla'_z(\nabla'_z)^k\varphi \ , \quad (\nabla''_{\bar z})^{k+1}\varphi = \nabla''_{\bar z}(\nabla''_{\bar z})^k\varphi \ .$$

It follows from Proposition 8.3 when ψ is horizontal and holomorphic that

$$(\nabla'_z)^p\varphi \in C^\infty(\psi) \ , \quad (\nabla''_{\bar z})^q\varphi \in C^\infty(\psi^\perp)$$

for all p and q, and hence that

$$(\nabla'_z)^p\varphi \perp (\nabla''_{\bar z})^q\varphi \ , \quad \forall\, p,q \geqslant 1 \ .$$

We say that a map $\varphi : M \to N$ which satisfies these last set of conditions is complex isotropic. Thus we have

Corollary 8.4 A horizontal and holomorphic map $\psi : M \longrightarrow G_k(T'N)$ gives rise to a complex isotropic harmonic map $\varphi : M \longrightarrow N$.

We would like to find a converse to this result which would allow us to construct a lift ψ out of φ. In this direction we define for each $\alpha \geq 1$

$$(W'_{\varphi,\alpha})_z = \mathrm{Span}_{\mathbb{C}}\{\nabla'_z\varphi, \ldots, (\nabla'_z)^{\alpha}\varphi\}$$

which is independent of the coordinates used to define it. Similarly define

$$(W''_{\varphi,\alpha})_z = \mathrm{Span}_{\mathbb{C}}\{\nabla''_{\bar{z}}\varphi, \ldots, (\nabla''_{\bar{z}})^{\alpha}\varphi\}.$$

$W'_{\varphi,\alpha}$, $W''_{\varphi,\alpha}$ are families of subspaces of $\varphi^{-1}T'N$ whose ranks may vary from point to point. As in section 6, in order to extend these families to subbundles over the set where they have lower rank we impose a curvature isotropy condition. We say φ is ∇' curvature isotropic if

$$R^N_o W'_{\varphi,\alpha} \subset W'_{\varphi,\alpha}, \quad \alpha = 1,2,\ldots.$$

The ∇' rank of φ is defined to be the maximum over z and α of the $(W'_{\varphi,\alpha})_z$. Suppose φ has ∇' rank k, than $W'_{\varphi,k}$ has rank k on a non-empty open set.

<u>Proposition 8.5</u> If $\varphi : M \to N$ is harmonic, ∇' curvature isotropic and of ∇' rank k then there is a lift φ'_k of φ to a map $M \to G_k(T'N)$ which agrees with $W'_{\varphi,k}$ where this has rank k.

<u>Proof</u> Centred at a point where the rank is less than k choose a holomorphic coordinate z and consider

$$\nabla'_z\varphi \wedge \ldots \wedge (\nabla'_z)^k\varphi \in C^\infty(\varphi^{-1}\Lambda^k T'N)$$

which is a smooth section vanishing at $z = 0$. By an argument analogous to the proof of Theorem 6.7 this section is holomorphic. It follows that the rank drops only at isolated points and that near such a point

$$\nabla'_z\varphi \wedge \ldots \wedge (\nabla'_z)^k \varphi = z^p s$$

for some zero-free local holomorphic section s of $\varphi^{-1}\Lambda^k T'N$ which is decomposible and hence defines φ'_k at $z = 0$, and spans $W'_{\varphi,k}$ for $z \neq 0$. This proves the Proposition.

Theorem 8.6 If $\varphi : M \to N$ is a harmonic, complex isotropic, ∇' curvature isotropic map of ∇' rank k then $\varphi'_k : M \to G_k(T'N)$ is horizontal and holomorphic.

Proof By construction $W'_{\varphi,k}$ is ∇'-stable, and is ∇''-stable since φ is harmonic and ∇' curvature isotropic. Since $W'_{\varphi,k}$ and φ'_k agree except at isolated points, it follows $C^\infty(\varphi'_k)$ is $(\varphi^{-1}\nabla)_X$-stable for all vector fields X, so φ'_k is horizontal as a map. By construction $\nabla'\varphi \in C^\infty(\varphi'_k)$, whilst complex isotropy implies $\nabla''\varphi \perp W'_{\varphi,k}$ so $\nabla''\varphi \perp \varphi'_k$ and hence φ'_k is holomorphic by Proposition 8.3.

In order to get a one-one correspondence we need notions of rank and curvature isotropy for maps into $G_k(T'N)$. These we obtain using the connection D on $J(N,h)$ which gives a connection on $G_k(T'N)$ which we also denote by D. For $\psi : M \to G_k(T'N)$ we let

$$d\psi(\partial/\partial z) = D'_z\psi + \overline{D''_z\psi}$$

be the decomposition into types relative to J_1 say, and

$$D'_z = (\psi^{-1}D)_{\partial/\partial z}\,, \quad D''_{\bar z} = (\psi^{-1}D)_{\partial/\partial\bar z}\,.$$

The D' rank of ψ is defined to be the maximum rank of the span of $D'_z\psi$, $(D'_z)^2\psi, \ldots$ as z varies. We let

$$(W'_{\psi,\alpha})_z = \mathrm{Span}_{\mathbb{C}}\{D'_z\psi, \ldots, (D'_z)^\alpha \psi\}$$

and say ψ is D' curvature isotropic if these subspaces are stable under the curvature operator

$$\psi^* R^D(\partial/\partial z, \partial/\partial \bar{z}).$$

Just as in section 6 we have the

Theorem 8.7 There is a one-one correspondence between horizontal and holomorphic D' curvature isotropic maps $\psi : M \to G_k(T'N)$ of D' rank k and the harmonic complex isotropic ∇' curvature isotropic maps $\varphi : M \to N$ of ∇' rank k given by $\varphi = \pi \circ \psi$, $\psi = \varphi'_k$.

Remark 8.8 Constant holmorphic sectional curvature for N makes harmonic complex isotropic maps ∇' curvature isotropic and this is the only situation where such maps are known. This was the case investigated by Eells and Wood in [5].

9. LOCAL AND GLOBAL LIFTS TO F-TWISTOR SPACES

We have seen in sections 6 and 8 that a non-constant conformal harmonic map $\varphi : M \longrightarrow N$ lifts to give a map $\varphi : M \rightarrow F_1(N,h)$ which is $\mathcal{F}_2$-holomorphic, and under suitable isotropy assumptions we get $\mathcal{F}_2$-holomorphic lifts to other f-twistor spaces $F_k(N,h)$, $G_k(T'N)$. It is possible to show, for many f-twistor spaces, that local lifts exist without any isotropy assumptions. This is purely an existence result and does not show how to construct such a lift. Salamon in his paper in this volume showed this first (for even dimensional N) for $J(N,h)$, and the extensions of this result we obtain in this section were suggested by J. Eells. When the domain is $\mathbb{C}P^1$ we are able to get some results concerning global lifts.

The first step in Salamon's proof is to study the set of complex structures on TN which make φ holomorphic. We replace this by

$$(9.1)\ F_k(\varphi) = \{(x,F) \in M\times F_k(N,h) \ . \ \varphi(x) = \pi(F),\ d\varphi_x \circ J^M_x = F \circ d\varphi_x\}$$

<u>Definition 9.1</u> If V is an inner product space, $\ell < k \leqslant \frac{1}{2} \dim V$ and $F' \in F_\ell(V)$, $F'' \in F_k(V)$ then we say $F' \subset F''$ if $V^{+,F'} \subset V^{+,F''}$.

We denote by $F_{\ell,k}(V)$ the set of pairs (F',F'') with $F' \in F_\ell(V)$, $F' \in F_k(V)$ and $F' \subset F''$. We have maps

$$\begin{array}{ccc} & F_{\ell,k}(V) & \\ {}^{\sigma_\ell}\swarrow & & \searrow^{\sigma_k} \\ F_\ell(V) & & F_k(V) \end{array}$$

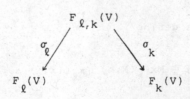

which allow us to embed $F_{\ell,k}(V)$ in $F_\ell(X) \times F_k(V)$ in the obvious way. It is easily seen that the orthogonal group $0(V)$ acts transitively on $F_{\ell,k}(V)$, the maps σ_ℓ, σ_k are equivariant, and the embedding of $F_{\ell,k}(V)$ makes it into a complex homogeneous submanifold of $F_\ell(V) \times F_k(V)$. This gives $F_{\ell,k}(V)$ an invariant complex structure for which σ_ℓ and σ_k are holomorphic maps.

Obviously this can be applied fibrewise to TN, to give a bundle $F_{\ell,k}(N,h)$ over N with maps

$$\begin{array}{ccc} & F_{\ell,k}(N,h) & \\ {}^{\sigma_\ell}\swarrow & & \searrow^{\sigma_k} \\ F_\ell(N,h) & & F_k(N,h) \end{array}$$

which are holomorphic on each fibre and preserve the horizontal distributions induced by the Levi-Civita connection. It follows that $F_{\ell,k}(N,h)$ can be given two f-twistor space structures using either of the maps σ_ℓ, σ_k for the map f in (7.1). We shall choose the map σ_ℓ and take the corresponding F_2 structure on $F_{\ell,k}(N,h)$. Then σ_ℓ is an f-morphism, and σ_k at least preserves the $+i$ eigenspaces, which we refer to as being F_2-holomorphic.

Suppose

$$\psi : M \longrightarrow F_\ell(N,h)$$

is a smooth map then we define

$$\widetilde{F}_k(\psi) = \{(x,F'') \in M \times F_k(N,h) \quad \pi_\ell \psi(x) = \pi_k F'',\ \psi(x) \subset F''\}.$$

Note that for φ a conformal harmonic map from M to N

$$F_k(\varphi) = \widetilde{F}_k(\widetilde{\varphi}) ,$$

where $\widetilde{\varphi} . M \longrightarrow F_1(N,h)$ is defined by Theorem 6.5.

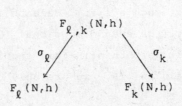

If X is a manifold with an f-structure F and Y is a submanifold of X we say Y is an f-submanifold if TY is stable under F. Obviously Y then has an induced f-structure for which the inclusion map is an f-morphism. If the induced f-structure on Y is an almost complex structure we say Y is an almost complex submanifold.

Theorem 9.2 If $\psi : M \to F_\ell(N,h)$ is an F_2-holomorphic map of the complex manifold M then $\widetilde{F}_k(\psi)$ is an almost complex submanifold of $M{\times}F_k(N,h)$.

Proof The graph Γ^ψ of $\varphi \subset M{\times}F_\ell(N,h)$ is obviously an almost complex submanifold. If we pull back the bundle $\sigma_\ell : F_{\ell,k}(N,h) \longrightarrow F_\ell(N,h)$ by ψ, then we get a submanifold of $M{\times}F_{\ell,k}(N,h)$ which is $(\mathrm{id} \times \sigma_\ell)^{-1}\Gamma^\psi$, and since we have complex submanifolds as fibres this is obviously an almost complex subamanifold of $M{\times}F_{\ell,k}(N,h)$. But

$$\widetilde{F}_k(\psi) = (\mathrm{id} \times \sigma_k)\{(\mathrm{id} \times \sigma_\ell)^{-1}\Gamma^\psi\}$$

and since σ_k preserves $+i$ eigenspaces of the F_2-structures, it follows that $\widetilde{F}_k(\psi)$ is an almost complex submanifold of $M{\times}F_k(N,h)$.

Theorem 9.3 If M is a Riemann surface and $\psi : M \longrightarrow F_\ell(N,h)$ is an F_2-holomorphic map then the induced almost complex structure on $\widetilde{F}_k(\psi)$ is integrable.

Proof The inclusion map

$$\imath : \widetilde{F}_k(\psi) \longrightarrow M{\times}F_k(N,h)$$

maps the almost complex structure on $\widetilde{F}_k(\psi)$ to the F_2-structure on $M{\times}F_k(N,h)$. It follows their Nijenhuis tensors are $\imath$-related, so if we can show that the tangents to $\widetilde{F}_k(\psi)$ are isotropic for N^{F_2}, then the almost complex structure on $\widetilde{F}_k(\psi)$ will be integrable. It suffices to take pairs of vectors in the $-\imath$ eigenspace which are tangent to $\widetilde{F}_k(\psi)$. Since the complex structure of M is integrable, it is easy to see that if (X,Y), (X',Y') are tangents to $M\times F_k(N,h)$ in the $-i$ eigenspace then

$$N^{F_2}((X,Y),(X',Y')) = N^{F_2}(Y,Y')$$

$$= [F_2Y, F_2Y'] - F_2[F_2Y,Y'] - F_2[Y, F_2Y'] + F_2^{\,2}[Y,Y']$$

$$= -[Y,Y'] + 2i F_2[Y,Y'] + F_2^{\,2}[Y,Y']$$

$$= (F_2 + i)^2 [Y,Y'] .$$

Now F_2 is parallel with respect to the connection D, and so

$$(F_2 + i)^2[Y,Y'] = -(F_2 + i)^2 T^D(Y,Y')$$

where T^D is the torsion of D which we calculated in section 4 Since ∇ is torsion-free,

$$d\pi \circ T^D = -P \wedge d\pi$$

and

$$[P \circ T^D, F] = [\pi^*R - \tfrac{1}{2}[P \wedge P], F] .$$

Now if both Y and Y' are in the $-i$ eigenspace of F_2 then $P(Y)$ and $P(Y')$ are both in the $+i$ eigenspace of J^V and so is $[P(Y),P(Y')]$. Also $d\pi(Y)$ and $d\pi(Y')$ are proportional since they are both in $d\varphi(T''_xM)$ where $\varphi = \pi_\ell \circ \psi$. Thus $\pi^*R(Y,Y') = 0$ and so

$$(F_2 + i)^2 P(T^D(Y,Y')) = -(-J^V + i)^2[P(Y),P(Y')] = 0.$$

This shows the vertical part is zero. On the other hand we have the horizontal part determined by

$$d\pi N^{F_2}(Y,Y') = -(F + i)^2\{P(Y)d\pi(Y') - P(Y')d\pi(Y)\}.$$

Now

$$(F + i)^2 P(Y)d\pi(Y') = (F + i)[F + i, P(Y)]d\pi(Y')$$

$$= -(F + i)(F^{-1}\nabla)_Y \; d\pi(Y')$$

so

$$d\pi N^{F_2}(Y,Y') = -(F + i)\{(\pi^{-1}\nabla)_Y F d\pi(Y') - (\pi^{-1}\nabla)_{Y'} F d\pi(Y)\}.$$

But if $\varphi = \pi_\ell \circ \psi$, then ψ F_2-holomorphic implies $\psi(x)d\varphi(X) = -id\varphi(X)$, so $Fd\varphi(X) = -id\varphi(X)$. Then $\widetilde{F}_k(\psi) \subset F_k(\varphi)$. But differentiating the definition of $F_k(\varphi)$,

$$T_{(x,F)}F_k(\varphi) = \{(X,Y) \quad d\varphi(X) = d\pi(Y),$$

$$d\varphi(X,JU) = (\pi^{-1}\nabla)_Y F d\varphi(U) + F\nabla d\varphi(X,U)$$

$$\forall\, U \in T_x M\}.$$

Thus

$$(\pi^{-1}\nabla)_Y F d\pi(Y') = (\pi^{-1}\nabla)_Y F d\pi(X')$$

$$= \nabla d\varphi(X,JX') - F\nabla d\varphi(X,X')$$

$$= -(F + i)\nabla d\varphi(X,X').$$

Interchanging (X,Y) with (X',Y') and subtracting gives

$$d\pi N^{F_2}(Y,Y') = 0$$

since $\nabla d\varphi$ is symmetric. Thus both horizontal and vertical parts of the Nijenhuis tensor vanish and hence the almost complex structure on $\widetilde{F}_k(\psi)$ is integrable as claimed.

Corollary 9.4 If $\varphi : M \longrightarrow N$ is a conformal harmonic map then φ can be locally lifted to an F_2-holomorphic map into $F_k(N,h)$ for any k.

Proof If φ is conformal harmonic then we have the F_2-holomorphic map $\tilde{\varphi} : M \longrightarrow F_1(N,h)$. Hence we get the complex manifold $\tilde{F}_k(\tilde{\varphi}) = F_k(\varphi)$ with the holomorphic submersion $F_k(\varphi) \longrightarrow M$. Any local holomorphic section of this map projects to $F_k(N,h)$ to give an F_2-holomorphic map lifting φ.

If N is even dimensional and $\dim N = 2k$ then $J(N,h) = F_k(N,h)$ and the Corollary then gives Salamon's result.

In Theorem 9.2 and 9.3 we can replace F_2 by F_1 and the results still hold.

These results extend also to immersed f-twistor spaces $f : Z \longrightarrow F_k(N,h)$:

Theorem 9.5 If $\varphi : M \longrightarrow N$ is a conformal harmonic map and $f : Z \longrightarrow F_k(N,h)$ an immersed f-twistor space then

$$Z_\varphi = \{(x,z) \in M\times Z : \varphi(x) = \pi(z), d\varphi_x \circ J_x = f(z) \circ d\varphi_x\}$$

is an almost complex submanifold of $M\times Z$ and this almost complex structure is integrable. Hence there are locally F_2-holomorphic lifts $\psi : U \longrightarrow Z$ of φ.

Proof Z_φ is immersed in $F_k(\varphi)$ by $(x,z) \longrightarrow (x,f(z))$ and is clearly an almost complex submanifold of $F_k(\varphi)$ so the results follow from those for $F_k(\varphi)$.

It is natural to ask if there are global lifts which are F_2-holomorphic. Salamon has pointed out that this is so for N of dimension 4 or 6 since $J(N,h)$ is a bundle of projective spaces in these dimensions. We shall show that for M a projective line $\mathbb{C}P^1$ we get global lifts to $F_k(N,h)$ for all k.

The method used is based on the bundle point of view. A map $\psi : M \to F_k(N,h)$ is F_2-holomorphic if, for $\varphi = \pi \circ \psi$, we have

(i) $\tilde{\varphi} \subset \psi^+$,

(ii) $(\varphi^{-1}\nabla)_X C^\infty(\psi^+) \subset C^\infty(\psi^+)$, $X \in C^\infty(T''M)$.

Condition (ii) means ψ^+ is a holomorphic subbundle of $\varphi^{-1}TN^{\mathbb{C}}$ for its Koszul-Malgrange holomorphic structure.

If we have a holomorphic bundle E, then $H^p(E)$ denotes the pth cohomology group of its sheaf of holomorphic sections and $h^p(E)$ the dimension of this group. We need the following Lemma:

<u>Lemma 9.6</u> If E is a holomorphic vector bundle on a Riemann surface M with a holomorphic bilinear form $(\ ,\ )$ and $h^o(E) \geqslant 2$ then E has an isotropic line subbundle L.

<u>Proof</u> For $s_1, s_2 \in H^o(E)$, (s_1, s_2) is holomorphic, so constant and hence defines a bilinear form on $H^o(E)$. Since $h^o(E) \geqslant 2$, this bilinear form has a non-zero isotropic vector s. Thus there is a non-zero holomorphic section s of E with $(s,s) = 0$. Then s determines a holomorphic line subbundle $L \subset E$ with $(L,L) = 0$.

<u>Proposition 9.7</u> If $\psi_k : \mathbb{C}P^1 \to F_k(N,h)$ is F_2-holomorphic and $2(k+1) \leqslant \dim N$ then there is $\psi_{k+1} : \mathbb{C}P^1 \to F_{k+1}(N,h)$ with $\psi_k^+ \subset \psi_{k+1}^+$ and ψ_{k+1} is F_2-holomorphic.

<u>Proof</u> Since ψ_k is F_2-holomorphic, $\tilde{\varphi} \subset \psi_k^+$ and ψ_k^+ is a holomorphic isotropic subbundle of $\varphi^{-1}TN^{\mathbb{C}}$ using the bilinear form $\varphi^{-1}h$ (which is holomorphic since it is parallel). Consider

$$\psi_k^\perp = \{X \in \varphi^{-1}TN^{\mathbb{C}} : \varphi^{-1}h(\psi_k^+, X) = 0\},$$

then $\psi^{\perp} \subset \varphi^{-1}TN^{\mathbb{C}}$ is holomorphic and $\varphi^{-1}h$ restricted to $\psi_k^{\perp}$ has ψ_k^{+} as its kernel so descends to give a non-singular holomorphic bilinear form $(\ ,\ )$ on $E = \psi_k^{\perp}/\psi_k^{+}$. Now consider Riemann-Roch for E:

$$h^{o}(E) - h^{1}(E) = c_1(E) + rk(E).$$

We have $rk(E) = n-2k \geqslant 2$ and $E \cong E^*$ so $c_1(E) = 0$ and hence $h^{o}(E) = h^{1}(E) + rk(E) \geqslant 2$. Thus by Lemma 9.6 we have a holomorphic isotropic line subbundle $L \subset E$ which lifts back to give $\psi_{K+1}^{+} \supset \psi_k^{+}$ which is a holomorphic and isotropic subbundle of $\varphi^{-1}TN^{\mathbb{C}}$.
Moreover $\tilde{\varphi} \subset \psi_k^{+} \subset \psi_{k+1}^{+}$ so ψ_{k+1} as a map to $F_{k+1}(N,h)$ is F_2-holomorphic.

<u>Theorem 9.8</u> A map $\varphi : \mathbb{C}P^1 \to N$ is conformal and harmonic if and only if for any k φ admits an F_2-holomorphic lift $\psi : \mathbb{C}P^1 \to F_k(N,h)$.

<u>Proof</u> If φ is conformal and harmonic then we have $\tilde{\varphi} : \mathbb{C}P^1 \to F_1(N,h)$ and so by applying Proposition 9.7 $k-1$ times we get
$\tilde{\varphi} = \psi_1^{+} \subset \psi_2^{+} \subset \ldots \subset \psi_k^{+}$ with $\psi_k : \mathbb{C}P^1 \to F_k(N,h)$ F_2-holomorphic.
The converse is of course a consequence of Theorem 5.6

<u>Remark 9.9</u> As pointed out by Salamon, this method can be applied locally to prove the Corollary 9.4, but does not give Theorem 9.5 so easily.

An alternative approach is to observe that an isotropic meromorphic section would also give us an isotropic line subbundle by treating the poles in the same way as we treated the zeros in Remark 6.4. If $E \longrightarrow M$ is a holomorphic vector bundle over a closed Riemann surface and s,t are meromorphic sections the (s,t) is a meromorphic function which we can no longer suppose is constant as we did in Lemma 9.6. Instead we consider $H^{o}(m(E))$ the space of global meromorphic sections which is a vector space over the meromorphic functions $m(M)$ on M of dimension equal to the rank of E. The holomorphic bilinear form on E gives an $m(M)$-valued quadratic form on $H^{o}(m(E))$ and we want a non-trivial zero of this. Since $m(M)$ is a function field in one variable over $\mathbb{C}$, a result of Lang and Tsen [7] (we thank

M. Reid for this reference) tells us that for rank $E \geq 3$ such a zero exists. Thus we can argue as in Proposition 9.7 to successively extend a given F_2-holomorphic map $\psi_k : M \to F_k(N,h)$ to ψ_{k+1} provided $n-2k \geq 3$. If n is odd we thus can go all the way from $\psi_1 = \tilde{\varphi}$ to a maximal isotropic subbundle ψ_q with $n = 2q+1$ and produce a flag $\tilde{\varphi} \subset \psi_2 \subset \ldots \subset \psi_2$ of F_2-holomorphic lifts. For n even we cannot immediately make the last step. Given a holomorphic bundle E of rank 2 we are looking for an isotropic line bundle, which is thus a section of the quadric bundle in $P(E)$. In rank 2 the fibres of this quadric bundle are pairs of points, so the total space is a double cover of M. It may thus be necessary to pass to a double covering of M to get an F_2-holomorphic lift to $J(N,h)$.

<u>Theorem 9.10</u> A map $\varphi \cdot M \longrightarrow N$ of a closed Riemann surface M which is conformal and harmonic admits an F_2-holomorphic lift $\psi . M \to F_k(N,h)$ for every k with $2k < \dim N$. If $\dim N = 2k$ and M is not simply connected an F_2-holomorphic lift to $J(N,h)$ always exists on a double covering of M.

<u>Remark 9.11</u> The lifting to a double covering in the above Theorem reflects the fact that we have made no assumption as to the orientability of N. If N is orientable it is possible to see that when we reach real codimension 4 we can in fact find two lifts as are found by Eells and Salamon [4] in their work on maps into oriented 4-manifolds. Thus for oriented N F_2-holomorphic lifts exist to all the f-twistor bundles.

10. THE LEVI-CIVITA CONNECTION AND CONDITION A ON $F_k(N,h)$

So far we did not consider a metric on $F_k(N,h)$. We can get one by taking invariant metric on $F_k(V,h_o)$ which then transports to each fibre to give a metric on V and then we lift the metric h on N horizontally. This metric makes π a Riemannian submersion and the connection D is metric. D is not the Levi-Civita connection as it has torsion T^D which we calculated in section 4. By subtracting off the torsion of D we are able to obtain the Levi-Civita connection $\hat{\nabla}$. If A denotes the $\mathrm{End}(TF_k(N,h))$-valued 1-form which is the difference of $\hat{\nabla}$ and D:

$$\hat{\nabla}_X = D_X + A(X),$$

then since $\hat{\nabla}$ is torsion-free we have

$$0 = T_D(X,Y) + A(X)Y - A(Y)X$$

from which by the standard argument we can solve for A. If the metric on $F_k(N,h)$ is denoted by $(\ ,\ )$ then

$$2(A(X)Y,Z) = -(T^D(X,Y),Z) + (T^D(Y,Z),X) - (T^D(Z,X),Y).$$

If the vertical and horizontal parts of a vector X on $F_k(N,h)$ are denoted by X^v and X^h respectively then

$$2(A(X)Y,Z) = 2(X^vY^h,Z^h) - (T^D(X,Y)^v,Z^v) + T^D(Y,Z)^v,X^v) - (T^D(Z,X)^vY^v),$$

where

$$T^D(X,Y)^v = P(T^D(X,Y)) = \{\pi^*R(X,Y) - [P(X),P(Y)]\}^m ,$$

by Proposition 4.8. This determines the Levi-Civita connection.

As a first application we calculate the second fundamental form $\nabla d\pi$ of π. It has values in $\pi^{-1}TN = E \cong H$, so we may assume Z is horizontal, then

$$\begin{aligned}
2(\nabla d\pi(X,Y),Z) &= 2((\pi^{-1}\nabla)_X d\pi(Y) - d\pi(\hat{\nabla}_X Y),Z) \\
&= 2(\{D_X + P(X)\}d\pi(Y) - d\pi(\hat{\nabla}_X Y),Z) \\
&= 2(d\pi((D_X - \hat{\nabla}_X)Y) + P(X)d\pi(Y),Z) \\
&= 2(-A(X)Y + X^v Y^h,Z) \\
&= -(T^D(Y,Z)^v,X^v) + (T^D(Z,X)^v,Y^v) \\
&= -(\pi^*R(Y^h,Z)^m,X^v) + (\pi^*R(Z,X^h)^m,Y^v).
\end{aligned}$$

Thus we have

<u>Theorem 10.1</u> The second fundamental form of the projection π is giver by

$$2(\nabla d\pi(X,Y),Z) = -(\pi^*R(Y^h,Z)^m,X^v) + (\pi^*R(Z^h,X^h)^m,Y^v)$$

for horizontal vectors Z. In particular $\nabla d\pi(X,Y)$ is zero if X and Y are both horizontal or both vertical, so π is harmonic.

We now determine when F_2 satisfies condition A. This is the condition (see section 2) that any holomorphic map of a Kaehler manifold into $F_k(N,h)$ is harmonic. F_2 will satisfy condition A if

$$\hat{\nabla}_X C^\infty(T^+F_k(N,h)) \subset C^\infty(T^+F_k(N,h)), \quad \forall\, X \in C^\infty(T^-F_k(N,h))$$

or equivalently

$$(10.1) \qquad (\hat{\nabla}_X Y, Z) = 0, \qquad \begin{array}{l} \forall\, X \in C^\infty(T^- F_k(N,h)), \quad Y \in C^\infty(T^+ F_k(N,h)), \\ Z \in C^\infty(T^o F_k(N, h)) + C^\infty(T^+ F_k(N,h)). \end{array}$$

This is equivalent to

$$0 = 2(D_X Y, Z) + 2(X^v Y^h, Z^h) - (T^D(X,Y)^v, Z^v) + (T^D(Y,Z)^v, X^v) - \\ - (T^D(Z,X)^v, Y^v).$$

Since D preserves F_2 and the $+i$ eigenspace is isotropic the first term is zero. For X in $T^- F_k(N,h)$ then X^v is in V' so corresponds with an element of m^+, whilst Y^h. Z^h correspond with elements of V^+ and $V^o + V^+$ respectively. Thus the second term is also zero. Thus condition A for F_2 becomes

$$0 = -(T^D(X,Y)^v, Z^v) + (T^D(Y,Z)^v, X^v) - (T^D(Z,X)^v, Y^v)$$

for X, Y, Z as in (10.1). Since $T^D(X,Y)$ vanishes unless X, Y are both horizontal or both vertical, if we take X vertical all the vectors in the above equation must be vertical, so we get the condition for the metric on the fibre to satisfy condition A. Taking X horizontal we get

$$0 = -(\pi^* R(X, Y^h)^m, Z^v) - (\pi^* R(Z^h, X)^m, Y^v)$$

and so R satisfies

$$(\pi^{-1} R(E^-, E^o + E^+)^m, V') = 0.$$

Since V' is maximally isotropic this implies

$$\pi^{-1} R(E^-, E^o + E^+)^m \subset V'.$$

Thus we have shown

Theorem 10.2 F_2 satisfies condition A if and only if the metric on the fibre satisfies condition A, and the curvature of N satisfies

$$\pi^{-1}R(E^-,E^o + E^+)^m \subset V''.$$

We can always take an invariant Kaehler metric on the fibre which therefore satisfies condition A, whilst the condition on R can be analyzed by representation theory as follows The fibre is an orbit of O(V), so as we move up and down the fibres of $F_k(N,h)$ we get each O(V) transform of the condition. Thus R takes values in the largest O(V) invariant subspace of the space of curvature tensors which satisfy

$$R(V^-,V^o + V^+)^m \subset m^-,$$

or

$$R(V^-,V^o + V^+) \subset h^{\mathbb{C}} + m^-.$$

If $\dim N \geq 5$, the curvature tensors decompose into Weyl tensors, traceless Ricci tensors and scalar curvature. It is not difficult to see that the scalar curvatures always satisfy this condition whilst the other types contain tensors which do not. Hence we have:

Theorem 10.3 If $\dim N \geq 5$, F_2 satisfies condition A on $F_k(N,h)$ if and only if the fibre metric satisfies condition A and h is Einstei and locally conformally flat.

In dimension 3 there are only two components of the curvature, the traceless Ricci tensors and the scalar curvature tensors and again only the scalar curvatures satisfy the condition. In this dimension the fibre metric automatically satisfies condition A, so we have

Theorem 10.4 If $\dim N = 3$, F_2 satisfies condition A on $F_1(N,h)$ if and only if N has constant curvature.

Finally, if N has dimension 4, since we are dealing with the full orthogonal group O(V) only the scalar curvature satisfies the condition. If N is oriented and we take $J_+(N,h)$ the complex structures on TN which induce the given orientation, then the fibre is a 2-sphere which satisfies condition A, whilst there are four irreducible parts to the curvature since the Weyl tensors split into self-dual and anti-self-dual parts of which only the anti-self-dual parts satisfy the condition. Thus we have

Theorem 10.5 If dim N = 4 then F_2 on $F_1(N,h)$ and $J(N,h)$ satisfies condition A if and only if N is Einstein and conformally flat. J_2 on $J_+(N,h)$ satisifes condition A if and only if N is Einstein and anti-self-dual.

Theorem 10.5 is due Eells and Salamon [4]. Note that anti-self-dual curvature is also the integrability condition for J_1.

11. CHARACTERISTIC CLASSES OF $F_k(N,h)$

The isomorphisms (3.1) apply fibrewise to the bundles V and E give

$$V' = \Lambda^2 E^+ + E^+ \otimes E^o,$$

$$V'' = \Lambda^2 E^- + E^- \otimes E^o.$$

Thus for F_2

$$T^+F_k(N,h) = E^+ + V'' = E^+ + \Lambda^2 E^- + E^- \otimes E^o .$$

If we define the Chern classes of F_2 to be those of $T^+F_k(N,h)$ then we have

$$c_1(F_2) = c_1(E^+) + c_1(\Lambda^2 E^-) + c_1(E^- \otimes E^o).$$

If $\dim N = n$, then

$$c_1(\Lambda^2 E^-) = -(k - 1)c_1(E^+),$$

$$c_1(E^- \otimes E^o) = -(n - 2k)c_1(E^+) + kc_1(E^o)$$

so

$$c_1(F_2) = -(n - k - 2)c_1(E^+) + kc_1(E^o).$$

Also

$$c_1(E^{\mathbb{C}}) = c_1(E^+) + c_1(E^-) + c_1(E^o) = c_1(E^o)$$

and $E^{\mathbb{C}} = \pi^{-1}TN^{\mathbb{C}}$ thus

$$c_1(E^{o}) = \pi^*\delta w_1(N)$$

where $w_1(N)$ is the first Stiefel-Whitney class of N and $\delta : H^1(N,\mathbb{Z}_2) \longrightarrow H^2(N,\mathbb{Z})$ is the Bockstein homomorphism. Thus

$$c_1(F_2) = -(n - k - 2)c_1(E^+) + k\pi^*\delta w_1(N).$$

Thus if k is even or N is oriented the second term is zero giving $c_1(F_2) = 0$ if $n = 4$, $k = 2$. In this case F_2 is the almost complex structure on the twistor space $J_+(N,h)$ of Eells and Salamon [4]. The only other case when $n - k - 2 = 0$ (since $n \geq 2k$) is $n = 3$, $k = 1$. Then

$$c_1(F_2) = \pi^*\delta w_1(N).$$

F_2 as a CR structure on $F_1(N,h)$ in this dimension has been studied by LeBrun [8].

Similar calculations can be performed for F_1 to give

$$c_1(F_1) = (n - k)c_1(E^+) + k\pi^*\delta w_1(N).$$

In their paper [4] Eells and Salamon introduced twistor degrees $d_\pm$ for a conformal harmonic map of a Riemann surface into an oriented 4-manifold. Let us consider the following more general situation of a smooth map

$$\psi : M \longrightarrow F_1(N,h).$$

For x in M $\psi(x)$ is thus an isotropic line in $T_{\varphi(x)}N^{\mathbb{C}}$ where φ is the composition $\pi \circ \psi : M \rightarrow N$. Such a line is equivalently an oriented 2-plane in $T_{\varphi(x)}N$ and we denote the orthogonal plane by $n_\psi(x)$ which also has a natural orientation and so gives a second isotropic line

$n'_\psi(x)$. We set

$$\psi_+ = \psi + n'_\psi, \quad \psi_- = \psi + \overline{n'_\psi}$$

which can be viewed as maps into $J_\pm(N,h)$, the bundles of complex structures on TN compatible with the metric and whose natural orientations agree or disgree with the given orientation of N respectively. If V' is the vertical (1,0) tangent bundle, which is a complex line bundle, then we set

$$D_\pm(\psi) = \langle c_1(\psi_\pm^{-1} V'),[M]\rangle = \psi_\pm^* \langle c_1(V'),[M]\rangle.$$

The twistor degrees are then defined by

$$d_\pm(\varphi) = \tfrac{1}{2} D_\pm(\varphi).$$

In general the twistor degrees will not be integers, but since $V' \cong \Lambda^2 E^+$, then $c_1(V') = c_1(E^+)$, so

$$c_1(\psi_\pm^{-1} V') = c_1(\psi) \pm c_1(n'_\psi)$$

viewing ψ and n'_ψ as subbundles of $\varphi^{-1}TN^{\mathbb{C}}$. Thus D_+ and D_- are always congruent mod 2 and so d_+ and d_- are simultaneously integral. Thus in discussing when the twistor degrees are integral it suffices to consider one case, say d_+.

In fact $D_+(\psi)$ is even if and only if $c_1(\psi_+^{-1}E^+)$ is congruent to zero mod. 2. But the mod 2 reduction of the first Chern class of a complex vector bundle is the second Stiefel-Whitney class w_2 of the underlying real bundle. But $\psi_+^{-1}E^+$ has underlying real bundle $\varphi^{-1}TN$ and so $D_+(\psi)$ is even if and only if $\varphi^* w_2(N)$ vanishes. In particular this is always the case if N is spin. We have thus shown

Proposition 11.1 The twistor degrees $d_\pm(\varphi)$ of a map $\varphi : M \longrightarrow N$ are integers if and only if $\varphi^* w_2(N) = 0$.

If we want a condition on N which guarantees that the twistor degrees of any conformal harmonic map are integers, then we have to demand that ν' on $J_+(N,h)$ is a square of a line bundle. This means that $\Lambda^2 E^+$ is a square which is equivalent to $\pi^{-1}TN$ being spin. It is not hard to show that $\pi^* : H^2(N,\mathbb{Z}_2) \longrightarrow H^2(J_+(N,h),\mathbb{Z}_2)$ is an injection, so $\pi^{-1}TN$ is spin if and only if TN is.

An example of a conformal harmonic map whose twistor degrees are not integers is the inclusion map $\mathbb{C}P^1 \xrightarrow{\ i\ } \mathbb{C}P^2$, which is in fact totally geodesic of degree one. $c_1(\tilde{i}) = 2$, whilst the normal degree is 1, so

$$D = 3, \quad D = 1 .$$

12. ISOMETRIC IMMERSIONS OF KAEHLER MANIFOLDS

Let (M,g,J) be almost Hermitian of real dimension $2m$ and (N,h) be a Riemannian manifold. If $\varphi : M \to N$ is an immersion we try to define a lift

$$\tilde{\varphi} : M \longrightarrow F_m(N,h)$$

by

$$\tilde{\varphi}(x)d\varphi(X) = d\varphi(J_x X), \quad X \in T_x M,$$

and

$$\tilde{\varphi}(x) = 0 \quad \text{on} \quad d\varphi(T_x M)^{\perp}.$$

From the definition of $F_m(N,h)$ we must have $\tilde{\varphi}(x)$ skew-symmetric which requires the vanishing of $(\varphi^*h)^{2,0}$. To guarantee this we shall assume that φ is isometric:

$$\varphi^*h = g$$

since g is (1.1) by assumption.

Thus let φ be an isometric immersion and denote by Q^{φ} the orthogonal projection onto $d\varphi(TM)$ in $\varphi^{-1}TN$. Then

$$Q^{\varphi} = -(F^{\tilde{\varphi}})^2 .$$

For an isometric immersion the Levi-Civita connections are related by

$$(\varphi^{-1}\nabla^N)_X d\varphi(Y) = d\varphi(\nabla^M_X Y) + \nabla d\varphi(X,Y)$$

and this is a decomposition into tangential and normal components. Thus

$$d\varphi(\nabla^M_X Y) = Q^\varphi(\varphi^{-1}\nabla^N)_X d\varphi(Y).$$

Since

$$F^{\tilde{\varphi}} = F^{\tilde{\varphi}} \circ Q^\varphi$$

we have

$$F^{\tilde{\varphi}}(\varphi^{-1}\nabla^N)\ d\varphi(Y) = F^{\tilde{\varphi}} d\varphi(\nabla^M_X Y) = d\varphi(J\nabla^M_X Y),$$

and hence

$$\{(\varphi^{-1}\nabla^N)_X F^{\tilde{\varphi}}\} d\varphi(Y) = (\varphi^{-1}\nabla^N)_X d\varphi(JY) - d\varphi(J\nabla^M_X Y)$$

$$= \nabla d\varphi(X,JY) + d\varphi((\nabla^M_X J)Y).$$

Thus we have proven the

Theorem 12.1 The lift $\tilde{\varphi} : M \to F_m(N,h)$ of an isometric immersion of an almost Hermitian manifold M into the Riemannian manifold N is horizontal if and only if M is Kaehler and φ totally geodesic.

In a series of papers [1] A.Adler studies isometric immersion of Kaehler manifolds into spheres and claims that the lift $\tilde{\varphi}$ (called g in this work) is always horizontal. It is clear from Theorem 12.1 that this is not generally so, and in the following we shall determine exactly what is the condition $\tilde{\varphi}$ must satisfy for M to be Kaehler. We have

$$d\varphi((\nabla^M_X J)Y) = d\varphi(\nabla^M_X(JY) - J\nabla^M_X Y)$$

$$= Q^{\varphi}(\varphi^{-1}\nabla^N)_X d\varphi(JY) - F^{\tilde{\varphi}}(\varphi^{-1}\nabla^N)_X d\varphi(Y)$$

$$= Q^{\varphi}(\varphi^{-1}\nabla^N)_X(F^{\tilde{\varphi}})d\varphi(Y).$$

Thus M is Kaehler if and only if

$$Q^{\varphi}(\varphi^{-1}\nabla^N)(F^{\varphi})Q^{\varphi} = 0.$$

Using

$$(\varphi^{-1}\nabla^N)F^{\tilde{\varphi}} = [\varphi*P, F^{\tilde{\varphi}}]$$

this is equivalent to

$$[Q^{\varphi}\tilde{\varphi}*PQ^{\varphi}, F^{\tilde{\varphi}}] = 0;$$

but $Q^{\varphi}\tilde{\varphi}*PQ^{\varphi}$ is in the image of $\mathrm{ad}F^{\tilde{\varphi}}$, hence M is Kaehler if and only if

$$Q^{\varphi}\tilde{\varphi}*PQ^{\varphi} = 0$$

if and only if

$$F^{\tilde{\varphi}}\varphi*PF^{\tilde{\varphi}} = 0.$$

We have shown:

Theorem 12.2 If $\tilde{\varphi}: M \longrightarrow F_m(N,h)$ is the lift of an isometric immersion where (M,g,J) is almost Hermitian, then M is Kaehler if and only if

$$\tilde{\varphi}*(FPF) = 0.$$

Thus FPF is a V-valued 1-form on $F_m(N,h)$ whose pull-back by $\tilde{\varphi}$ is the obstruction to M being Kaehler. $\tilde{\varphi}$ horizontal is the stronger condition $\tilde{\varphi}^*P = 0$. We can reformulate the condition $\tilde{\varphi}^*(FPF) = 0$ rather differently in terms of the connection D. Recall that

$$D = \pi^{-1}\nabla^N - P$$

so that

$$\varphi^{-1}D = \tilde{\varphi}^{-1}\nabla^N - \tilde{\varphi}^*P.$$

Note also that

$$DF = 0$$

implies

$$(\tilde{\varphi}^{-1}D)Q^{\varphi} = 0$$

and hence

$$d\varphi(\nabla^M_X Y) = Q^{\varphi}(\varphi^{-1}\nabla^N)_X d\varphi(Y)$$

$$(12.1) \qquad = Q^{\varphi}(\tilde{\varphi}^{-1}D)_X d\varphi(Y) + Q^{\varphi}(\tilde{\varphi}^*P)(X)d\varphi(Y)$$

$$= (\tilde{\varphi}^{-1}D)_X d\varphi(Y) + Q^{\varphi}(\tilde{\varphi}^*P)(X)d\varphi(Y).$$

Thus $Q^{\varphi}\tilde{\varphi}^*PQ^{\varphi} = 0$ if and only if $d\varphi(\nabla^M_X Y) = (\tilde{\varphi}^{-1}D)_X d\varphi(Y)$. Note that $\tilde{\varphi}^{-1}D$ stabilizes $d\varphi(TM) \subset \varphi^{-1}TN = \tilde{\varphi}^{-1}E$ and induces a connection on M for any almost Hermitian manifold. Denote this connection by D^{φ}.

Proposition 12.3

$$Q^{\varphi}\tilde{\varphi}^*P(X)d\varphi(Y) = \tfrac{1}{2}d\varphi(J(\nabla^M_X J)Y)$$

Proof From $J^2 = -1$ follows

$$J\nabla^M_X J + (\nabla^M_X J)J = 0$$

and hence

$$2J\nabla^M_X J = [J, \nabla^M_X J].$$

Thus

$$2d\varphi(J(\nabla^M_X J)Y) = d\varphi([J, \nabla^M_X J]Y)$$

$$= Q^{\varphi}[F^{\tilde{\varphi}}, (\varphi^{-1}\nabla^N)_X F^{\tilde{\varphi}}]d\varphi(Y).$$

But $\mathrm{ad}F^{\tilde{\varphi}}$ has eigenvalues $\pm i$, $\pm 2i$ on $\tilde{\varphi}^{-1}V$, so if we split $\varphi^*P(X)$ into its components in these eigenspaces,

$$\tilde{\varphi}^*P(X) = A_i + A_{-i} + A_{2i} + A_{-2i}$$

then

$$(\varphi^{-1}\nabla^N)_X F^{\tilde{\varphi}} = [\tilde{\varphi}^*P(X), F^{\tilde{\varphi}}] = -iA_i + iA_{-i} - 2iA_{2i} + 2iA_{-2i}$$

so

$$[(\varphi^{-1}\nabla^N)_X F^{\tilde{\varphi}}, F^{\tilde{\varphi}}] = -A_i - A_{-i} - 4A_{2i} - 4A_{-2i}.$$

Now Q^{φ} maps $\tilde{\varphi}^{-1}E^{\mathbb{C}}$ onto $\tilde{\varphi}^{-1}E^+ + \tilde{\varphi}^{-1}E^-$ and $A_{\pm i}$ maps these to $\tilde{\varphi}^{-1}E^o$ which is the kernel of Q^{φ}, thus

$$Q^{\varphi}[(\varphi^{-1}\nabla^N)_X F^{\tilde{\varphi}}, F^{\tilde{\varphi}}]Q^{\varphi} = -4Q^{\varphi}(A_{2i} + A_{-2i})Q^{\varphi}$$

whilst

$$Q^{\varphi}\tilde{\varphi}^*P(X)Q^{\varphi} = Q^{\varphi}(A_{2i} + A_{-2i})Q^{\varphi}$$

which proves the result.

We thus have

$$d\varphi(D^{\varphi}_X Y) = (\tilde{\varphi}^{-1}D)_X d\varphi(Y)$$

$$= d\varphi(\nabla^M_X Y) - Q^{\varphi}(\tilde{\varphi}^* P)(X)d\varphi(Y)$$

by (12.1) and so by the Proposition, since $d\varphi$ is injective,

$$D^{\varphi}_X Y = \nabla^M_X Y - \tfrac{1}{2}J\nabla^M_X JY.$$

Thus we have shown that M is Kaehler if and only if $D^{\varphi}_{\sim} = \nabla^M$, and that in this case $d\varphi$ gives an isomorphism of TM with $\mathrm{Im}F^{\varphi}$ which takes ∇^M to the connection induced on $\mathrm{Im}F^{\varphi}$ by $\tilde{\varphi}^{-1}D$. This allows us to find tensors on $F_m(N,h)$ which induce standard tensors on M by pull-back by $\tilde{\varphi}$. For example the Kaehler form of M is given by $\tilde{\varphi}^*\Omega$ where Ω is the horizontal 2-form

$$\Omega(X,Y) = \pi^{-1}h(d\pi(X), Fd\pi(Y)).$$

The first Chern form c_1 of M is given by the Ricci form

$$c_1(X,Y) = \frac{1}{2\pi i}\,\mathrm{Trace}_{T'M} R^M(X,Y).$$

Then for M Kaehler,

$$c_1 = \tilde{\varphi}^* C_1$$

where C_1 is the 2-form on $F_m(N,h)$ given by

$$C_1(X,Y) = \frac{1}{2\pi i}\,\mathrm{Trace}_{E^+}(R^D(X,Y))$$

where R^D is the curvature of D as a connection in E^+.

Other formulae of Adler may be obtained by using $\tilde{\varphi}^*D$ rather than $\varphi^{-1}\nabla^N$

Let $G_{2m}(N)$ be the Grassmann bundle of oriented 2m-planes in TN and

$$\rho : F_m(N,h) \longrightarrow G_{2m}(N)$$

the map sending F to its image with orientation given by the complex structure which F induces on its image. If $\varphi : M \to N$ is an immersion, then we have a lift

$$\hat{\varphi} : M \longrightarrow G_{2m}(N)$$

given by

$$\hat{\varphi}(x) = d\varphi(T_x M),$$

transporting the orientation of $T_x M$ to $\varphi(x)$. If φ is an isometric immersion the two lifts are related by

$$\hat{\varphi} = \rho \circ \tilde{\varphi}$$

Conversely, let M be a manifold and

$$\psi : M \longrightarrow F_m(N,h)$$

be an isometric immersion such that

(i) $\rho \circ \psi = (\pi \circ \psi)\hat{}$,

(ii) $d\psi(T_x M) \cap V_{\psi(x)} = 0, \quad \forall\, x \in M$

then the image of $d(\pi \circ \psi)_x$ is the image of $\psi(x)$ and so we get an endomorphism J_x of $T_x M$ such that

$$d(\pi \circ \psi)_x \circ J_x = \psi(x) \circ d(\pi \circ \psi)_x .$$

Giving M the pull-back metric makes M almost Hermitian with

$$\psi = \pi \circ \psi$$

and this induced almost Hermitian structure is Kaehler if and only if

$$\psi^*(FPF) = 0.$$

BIBLIOGRAPHY

[1] Adler A.: Classifying spaces for Kaehler metrics I-IV, Math. Annalen 152 (1963) 164-184, 154 (1964)257-266, 156 (1964) 378-392, 160 (1965) 41-58.

[2] Bryant R. Lie groups and twistor spaces. (Preprint).

[3] Eells J., Lemaire L.: A report on harmonic maps. Bull. Lond. Math. Soc. 10 (1978) 1-68.

[4] Eells J., Salamon S.: Twistorial construction of harmonic maps of surfaces into four-manifolds. (To appear).

[5] Eells J., Wood J.C.: Harmonic maps from surfaces to complex projective spaces. Advances in Math. 49 (1983) 217-263.

[6] Koszul J., Malgrange B.. Sur certaines structures fibrées complexes. Arch. math. 9 (1958) 102-109.

[7] Lang S.: On quasi algebraic closure. Annals of Math. 55 (1952) 373-390.

[8] LeBrun C.: The embedding problem for twistor CR manifolds. (Berkeley preprint).

[9] Lichnerowicz A.. Applications harmoniques et variétes kähleriennes. Symp. Math. III (1970) 341-402.

[10] O'Brian N., Rawnsley J.: Twistor spaces. Annals of Global Analysis and Geometry. (To appear).

[11] Salamon S.: Harmonic and holomorphic maps. (These Notes).

[12] Yano K.: on a structure defined by a tensor field f of type (1,1) satisfying $f^3 + f = 0$. Tensor 14 (1963) 99-109.

SIMON SALAMON

HARMONIC AND HOLOMORPHIC MAPS

INTRODUCTION

Suppose that $\varphi: U \longrightarrow V$ is a holomorphic mapping between open sets $U \subset \mathbb{C}^m$ and $V \subset \mathbb{C}^n$. Writing $\varphi = (\varphi^1, \ldots, \varphi^n)$, each component φ^i satisfies the system of Cauchy-Riemann equations

$$\frac{\partial \varphi^i}{\partial \bar{z}^j} = 0, \quad j = 1, \ldots, m.$$

Consequently φ is harmonic in the sense that each component satisfies Laplace's equation

$$\sum_{j=1}^{m} \frac{\partial^2 \varphi^i}{\partial z^j \partial \bar{z}^j} = 0.$$

More generally one can talk of holomorphic mappings between almost complex manifolds, and harmonic mappings between Riemannian manifolds. Putting the two together, a proposition of Lichnerowicz [Li] gives precise conditions on almost Hermitian manifolds M, N to ensure that a holomorphic map $\varphi: M \longrightarrow N$ is necessarily harmonic. The first two sections provide the background for this result.

In section 3 these ideas are generalized by associating to any even-dimensional Riemannian manifold N an almost complex manifold S which is a bundle over N with fibre $SO(2n)/U(n)$. The manifold S is a so-called twistor space in that it parametrizes complex structures on the tangent spaces of N, and itself comes equipped with two almost complex structures J_1 and J_2. The former is occasionally integrable, and for N 4-dimensional was used to classify self-dual solutions of the Yang-Mills equations [AHS]. On the other hand J_2 is never integrable, but does play a key role in the theory of harmonic maps (theorem 3.5).

Section 4 onwards is devoted mainly to a study of oriented minimal surfaces in N, i.e. conformal harmonic maps $\varphi : M \longrightarrow N$ where M is a Riemann surface. These are characterized in terms of J_2-holomorphic curves in S, ones whose differential is complex linear with respect to J_2. Other types of maps, namely totally umbilic and real isotropic ones, relate to J_1, and in this context we explain how Calabi used S to study minimal surfaces in a sphere S^{2n} [Ca_2].

Section 5 contains a description of various bundles T over Riemannian symmetric spaces, inspired by work of Bérard Bergery & Ochiai [BO] who formalized the twistor space concept using G-structures. An independent and more detailed treatment of this topic has been given by Bryant [Br_4]. Each T is a subbundle of S which is both a complex manifold and a 3-symmetric space, and gives rise to its own special class of minimal surfaces in N; we enumerate the possibilities using results of Wolf & Gray [WG].

The remaining three sections concentrate on examples of the above techniques, and provide a unification of work of various authors. After first discussing twistor spaces of complex Grassmannians, we use the J_2 formalism to define transforms of harmonic maps into complex projective space $\mathbb{C}P^n$. This provides an introduction to the study of almost complex structures on flag manifolds, which is a valuable tool for understanding minimal surfaces in symmetric spaces. Quaternionic Kähler manifolds are also considered since these have particularly simple twistor spaces.

The relationship with Gauss maps is discussed in section 7 by replacing the almost complex manifold S by a partially complex manifold $\tilde{G}_2(TN)$, namely a Grassmannian bundle defined over a Riemannian manifold N of arbitrary dimension. When N is 3-dimensional, this construction was used by LeBrun [Le] to create examples of nonrealizable codimension one CR manifolds. In fact S and $\tilde{G}_2(TN)$ belong to a whole family of twistor spaces studied by Rawnsley [Ra] in this volume. As a special example the above techniques are applied to the exceptional geometry of the sphere S^6.

When N is 4-dimensional, there is a close relationship between S and $\tilde{G}_2(TN)$, leading to some interesting correspondences (theorem 8.2). The theory is most rich when the curvature tensor of N is

Einstein and either self-dual or anti-self-dual. As a corollary we show that minimal surfaces in S^4 can be realized as minimal surfaces in $\mathbb{C}P^3$, somewhat in the spirit of the Penrose transform.

These notes represent an expanded version of seminars given at the Scuola Normale in 1983, and were completed at the Institute for Advanced Study. Accordingly the author acknowledges financial support from both the Consiglio Nazionale delle Ricerche and the National Science Foundation. Much of the work was carried out jointly with J. Eells. The first ideas appeared in [ES_1], and notation and preparatory material is taken from the paper [ES_2] which deals in more depth with the 4-dimensional case.

1. ALMOST HERMITIAN GEOMETRY

The most important class of manifolds for which the notion of a holomorphic mapping exists is the class of almost complex manifolds. A manifold M is said to be <u>almost complex</u> if there exists a section $J = J^M$ of End TM with $J^2 = -1$. All the objects defined and used in this work are supposed to be smooth. The tensor J is called the almost complex structure. The manifold M necessarily has even dimension. $2m$, and may be oriented by decreeing that whenever $\{X_1, JX_1, \ldots, X_m, JX_m\}$ is a basis of $T_x M$ then it is an oriented basis.

A mapping $\varphi : M \longrightarrow N$ between almost complex manifolds is said to be <u>holomorphic</u> if its differential $\varphi_* \; TM \rightarrow TN$ commutes with the respective almost complex structures, i.e.

$$\varphi_* \circ J^M = J^N \circ \varphi_*$$

The almost complex structure J of M is said to be <u>integrable</u>, and M is said to be a <u>complex manifold</u>, if locally there exist coordinates $z_r = x_r + iy_r$, $1 \leqslant r \leqslant m$, for which

$$J(\frac{\partial}{\partial x^r}) = \frac{\partial}{\partial y^r} .$$

Euclidean space $\mathbb{C}^m$ has a natural almost complex structure J with this property and holomorphic maps take on their usual meaning. More generally, M is complex iff each point has a neighbourhood admitting a holomorphic diffeomorphism into $\mathbb{C}^m$. However we emphasize that our use of the word holomorphic does not imply integrability.

The set of all frames of the form $\{X_1, JX_1, \ldots, X_m, JX_m\}$ determines a $GL(m,\mathbb{C})$-structure on M, i.e. a principal subbundle of the frame bundle with group $GL(m,\mathbb{C})$. This structure can always be reduced further to the maximal compact subgroup $U(m)$, which amounts to choosing a Riemannian metric g for which J is an orthogonal transformation.

The resulting manifold is then both Riemannian and almost complex with

$$g(JX,JY) = g(X,Y), \quad X,Y \in TM.$$

Such a manifold is called <u>almost Hermitian</u>, the term Hermitian being reserved for when J is integrable. For the remainder of this section M will denote an almost Hermitian manifold, with corresponding principal $U(m)$-bundle P.

The almost complex structure of M yields a decomposition

$$(TM)^{\mathbb{C}} = T^{1,0} \oplus T^{0,1}$$

of the complexified tangent bundle corresponding to the $\pm i$-eigenspaces of J. Taking the rth exterior power gives

$$(1.1) \qquad (\Lambda^r TM)^{\mathbb{C}} = \bigoplus_{p+q=r} T^{p,q},$$

where $T^{p,q} \cong \Lambda^p(T^{1,0}) \otimes \Lambda^q(T^{0,1})$. There are analogous definitions with TM replaced by T^*M, and we are using the notation $T^{p,q}$ rather than $\Lambda^{p,q}$ to emphasize that we are dealing with vectors. Indeed it is unnecessary to use forms at all, since there are natural isomorphisms

$$(1.2) \qquad (T^{p,q})^* \cong \overline{T^{p,q}} \cong T^{q,p},$$

the former induced from the metric g. Let $\{X_1,\ldots,X_m\}$ be an orthonormal basis of some tangent space T_xM, and extend g to the space $\Lambda^r T_xM$ by declaring $\{X_{i_1} \wedge \ldots \wedge X_{i_r},\ i_1 < \ldots < i_r\}$ to be an orthonormal basis. Then the sum (1.1) is orthogonal, and we can let g denote also its own complex extension to each $T^{p,q}$.

The vector bundle $T^{p,q}$ is really associated to P by means of a representation of $U(m)$, but this representation is not in general irreducible. To see this one considers the so-called fundamental 2-form or Kähler form ω on M, defined by

$$\omega(X,Y) = g(JX,Y), \quad X,Y \in TM.$$

In view of our preference for vectors though, we shall work instead with the "fundamental 2-vector" F dual to ω, so that

$$g(F, X \wedge Y) = g(JX,Y) = -g(X,JY).$$

Then F is a section of $T^{1,1}$, and there is an orthogonal splitting

$$T^{1,1} = \mathbb{R} \oplus T_0^{1,1}.$$

In fact if one defines the primitive subbundle $T_0^{p,q}$ of $T^{p,q}$ to be the orthogonal complement of $\{F \wedge \sigma : \sigma \in T^{p-1,q-1}\}$, one obtains

$$(1.3) \qquad T^{p,q} \cong \bigoplus_{r=0}^{\min(p,q)} T_0^{p-r,q-r},$$

with $T_0^{p,q} = 0$ if $p + q > m$. The representation of $U(m)$ corresponding to $T_0^{p,q}$ is known to be irreducible; this applies in particular to $T_0^{p,0} = T^{p,0}$.

The Levi-Civita connection

When choosing a connection on an almost Hermitian manifold M, there are two basic possibilities. One can either use the Levi-Civita connection ∇, i.e. the unique torsion-free connection with $\nabla g = 0$, or alternatively one can choose a connection $\overline{\nabla}$ on the principal $U(m)$-bundle P so that $\overline{\nabla} g = 0 = \overline{\nabla} J$. Because we shall be treating M primarily as a Riemannian manifold, we elect to work with ∇. Now any two of the tensors g,J,F determine the third, so in the presence of g, ∇J and ∇F carry the same information.

Lemma 1.1 For any $\alpha, \beta \in \Gamma(M, T^{1,0})$, $X \in (TM)^{\mathbb{C}}$,

$$g(\nabla_X F, \alpha \wedge \beta) = 2ig(\nabla_X \alpha, \beta)$$

$$g(\nabla_X F, \alpha \wedge \bar{\beta}) = 0.$$

Proof Using $\nabla g = 0$,

$$\begin{aligned} g(\nabla_X F, \alpha \wedge \beta) &= Xg(F, \alpha \wedge \beta) - g(F, \nabla_X \alpha \wedge \beta) - g(F, \alpha \wedge \nabla_X \beta) \\ &= iXg(\alpha, \beta) + ig(\nabla_X \alpha, \beta) - ig(\alpha, \nabla_X \beta) \\ &= 2ig(\nabla_X \alpha, \beta). \end{aligned}$$

Repeating the same procedure with $\bar{\beta}$ in place of β establishes the second equation. ∎

Lemma 1.1 relates $\nabla_X F$ with the second fundamental form η_X of the subbundle $T^{1,0}$ in $(TM)^{\mathbb{C}}$. The latter is the composition

(1.4) $$\eta_X : T^{1,0} \xrightarrow{\nabla_X} (TM)^{\mathbb{C}} \longrightarrow T^{0,1}$$

of the covariant derivative ∇_X with the projection, and measures the extent to which ∇_X fails to preserve $T^{1,0}$. Because of the projection η_X is actually a tensor, i.e. a section of

$$\mathrm{Hom}(T^{1,0}, T^{0,1}) \cong T^{0,1} \otimes T^{0,1}$$

(see (1.2)). Lemma 1.1 now shows that η_X belongs to the subspace $\Gamma^{0,2}$ of the right-hand side, and

(1.5) $$\nabla_X F = 2\,\mathrm{Re}(i\eta_X).$$

Using (1.2) again,

$$T^{0,2} \otimes (T^*M)^{\mathbb{C}} \cong T^{0,2} \otimes (T^{0,1} \oplus T^{1,0})$$

$$\cong (T^{0,1} \otimes T^{0,2}) \oplus T^{1,2}$$

Therefore it is possible to write

(1.6) $$\nabla F = D_1F + D_2F,$$

where

$$D_1F \in (T^{0,1} \otimes T^{0,2}) \oplus (T^{1,0} \otimes T^{2,0})$$

$$D_2F \in T^{1,2} \oplus T^{2,1}.$$

with these identifications, (1.4) yields (cf. [Ra, proposition 2.6])

Lemma 1.2 $D_1F = 0 \iff \nabla_X(T^{1,0}) \subset T^{1,0}, \quad \forall X \in \Gamma(M,T^{1,0});$

$D_2F = 0 \iff \nabla_X(T^{1,0}) \subset T^{1,0}, \quad \forall X \in \Gamma(M,T^{0,1})$

The tensors D_1F, D_2F represent very different properties of the almost Hermitian manifold M

Proposition 1.3 $D_1F = 0 \iff J$ is integrable;

$D_2F = 0 \iff (d\omega)^{1,2} = 0.$

Proof D_1F is determined by the tensor

$$\alpha \otimes \beta \otimes \gamma \longmapsto g(\nabla_\alpha \beta, \gamma), \quad \alpha,\beta,\gamma \in T^{1,0}.$$

By lemma 1.1, this tensor is skew in β,γ, so can be identified with its image

$$g(\nabla_\alpha \beta,\gamma) - g(\nabla_\beta \alpha,\gamma) = g([\alpha,\beta],\gamma)$$

under the isomorphism

$$V \otimes \Lambda^2 V \hookrightarrow \otimes^3 V \longrightarrow \Lambda^2 V \otimes V, \qquad V = (T^{1,0})^* .$$

Hence $D_1 F = 0$ iff $T^{1,0}$ is closed under Lie Bracket, which by the Newlander-Nirenberg theorem is precisely the condition that J be integrable. In fact putting $\alpha = X - iJX$, $\beta = Y - iJY$, $\gamma = Z - iJZ$ gives

$$\mathrm{Re}\; g([\alpha,\beta],\gamma) = g(N(X,Y),Z),$$

where

$$N(X,Y) = [X,Y] + J[JX,Y] + J[X,JY] - [JX,JY]$$

is the usual Nijenhuis tensor.

The (1.2)-component $(d\omega)^{1,2}$ of the 3-form $d\omega$ is given by

$$3d\omega(\alpha,\beta,\bar{\gamma}) = (\nabla_\alpha \omega)(\beta,\bar{\gamma}) + (\nabla_\beta \omega)(\bar{\gamma},\alpha) + (\nabla_{\bar{\gamma}}\omega)(\alpha,\beta),$$

where $\alpha,\beta,\gamma \in \Gamma(M,T^{1,0})$. The result concerning $D_2 F$ now follows from the fact that the first two terms on the right-hand side vanish in analogy with lemma 1.1, whereas the third equals $g(\nabla_{\bar{\gamma}} F, \alpha \wedge \beta)$. ∎

Types of almost Hermitian manifolds

An almost manifold M with $\nabla F = 0$ is said to be Kähler. In this case $T^{1,0}$ is completely preserved by the Levi-Civita connection which therefore reduces to a connection on the principal bundle P; the Riemannian holonomy group lies in $U(m)$. By proposition 1.3, to put a Kähler metric on a complex manifold it suffices to find a 2-form ω which is (1,2)-closed and such that $\omega(X,JY)$ is positive definite.

A (necessarily even-dimensional) manifold with a closed non-degenerate 2-form ω is called symplectic. Accordingly we shall call an almost Hermitian manifold with $D_2F = 0$ <u>(1,2)-symplectic</u>; this is in some sense the "opposite" of being a complex manifold. The next fact was observed in [FI], and will be pursued in section 7.

<u>Proposition 1.4</u> The sphere S^6 has a natural (1,2)-symplectic structure.

<u>Proof</u> The exceptional Lie group G_2 has an orthogonal representation on $\mathbb{R}^7$ inducing a transitive action on S^6 with isotropy subgroup SU(3). The reductive Lie algebra

$$g_2 = su(3) \oplus m, \quad [su(3),m] \subset m$$

gives rise to an SU(3)-connection $\overline{\nabla}$ on $S^6 = G_2/SU(3)$ whose torsion τ can be identified with the m-component of $[m,m]$. The SU(3)-structure makes S^6 into an almost Hermitian manifold whose metric is the standard Euclidean one induced from $\mathbb{R}^7$. If ∇ and F have their usual meanings, the quantity $\nabla - \overline{\nabla}$ is a tensor which can be identified with τ, and $\nabla F = (\nabla - \overline{\nabla})F$ defines an SU(3)-invariant component of τ.

Under U(m) there are decompositions

$$T^{0,1} \otimes T^{0,2} \cong T^{0,3} \oplus A$$

(1.7)

$$T^{1,2} \cong T^{0,1} \oplus T_0^{1,2}$$

(cf. (1.3)), in which A is irreducible. The only invariants under SU(3) in the above spaces lie in the trivial 1-dimensional representation $T^{0,3}$; thus for S^6 one must have

$$D_1F \in T^{3,0} \oplus T^{0,3}$$

(1.8)

$$D_2F = 0.$$

It follows that not only does $(d\omega)^{1,2}$ vanish, but in addition $(d\omega)^{0,3}$ completely determines ∇F. ∎

Almost Hermitian manifolds, Shading = Kähler

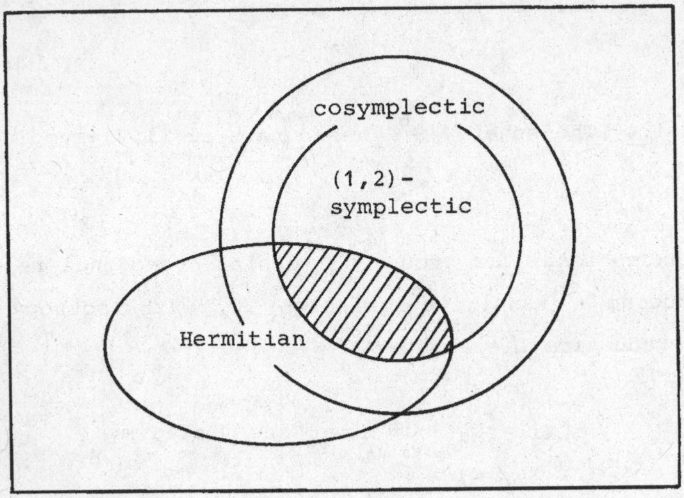

A complete classification of the various classes of almost Hermitian manifolds that can be defined using (1.7) has been given by Gray & Hervella [GH]. We mention one other class that will be important in the sequel. If tr denotes the U(m)-homomorphism $\Lambda^3 T \to T$ consisting of contraction with ω, it is easy to verify that (up to a constant) $tr(D_2 F)$ is dual to $d^*\omega$, where $d^* = -{*}d{*}$ is the codifferential. An almost Hermitian manifold with $tr(D_2 F) = 0$ is said to be cosymplectic; this is weaker than being (1,2)-symplectic.

The expressions almost-Kähler, quasi-Kähler or *0, semi-Kähler or almost semi-Kähler, nearly-Kähler are sometimes used in place of respectively symplectic, (1,2)-symplectic, cosymplectic, S^6-like. However we shall use the latter terminology because it causes less confusion when translated. In the diagram, the three regions represent the classes of Hermitian, (1,2)-symplectic, and cosymplectic manifolds.

2. HARMONIC MAPS AND LICHNEROWICZ'S PROPOSITION

Let $\varphi\colon M \to N$ be a smooth mapping between any two Riemannian manifolds with metrics g,h respectively. The differential $\varphi_*\colon TM \to TN$ can be interpreted as a homomorphism from the tangent bundle TM of M to the pullback $\varphi^{-1}TN$ of the tangent bundle of N. This homomorphism, which we denote by $d\varphi$, has the advantage of relating vector bundles defined over the same manifold M:

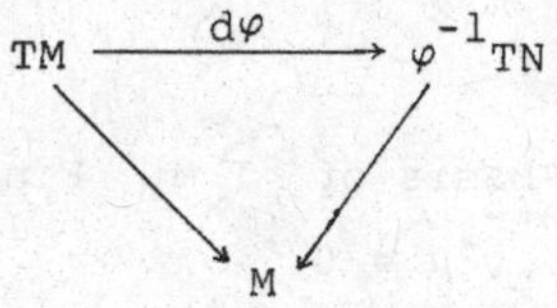

Both TM and TN admit a Levi-Civita connection; let ∇ denote both the connection on TM and the pullback on $\varphi^{-1}TN$. The covariant derivative of $d\varphi$ is then defined by

$$(\nabla_X d\varphi)(Y) = \nabla_X(\varphi_* Y) - \varphi_*(\nabla_X Y), \quad X,Y \in \Gamma(M,TM),$$

where on the right-hand side the fibres of TN and $\varphi^{-1}TN$ have been identified. Using the torsion-free property of the connections,

$$(\nabla_X d\varphi)(Y) - (\nabla_Y d\varphi)(X) = [\varphi_* X, \varphi_* Y] - \varphi_*[X,Y] = 0.$$

In other words $\nabla d\varphi \in \Gamma(M, S^2T^*M \otimes \varphi^{-1}TN)$ is a symmetric bilinear form on TM with values in $\varphi^{-1}TN$.

Suppose for the moment that φ is a Riemannian immersion, i.e. $\varphi^*h = g$. Then for any vector field X on M,

$$\nabla(\varphi_* X) = \varphi_*(\nabla X) + \nabla^{\perp}(\varphi_* X),$$

where $\nabla^{\perp}$ denotes the covariant derivative on $\varphi^{-1}TN$ followed by a projection to the orthogonal complement $(\varphi_* TM)^{\perp}$ of $\varphi_* TM$ in TN.

It follows that

$$(\nabla d\varphi)(X) = \nabla^{\perp}(\varphi_* X).$$

and $\nabla d\varphi$ is the second fundamental form of the subbundle $\varphi_* TM$ in TN (cf. (1.4)); this is of course the classical second fundamental form of the submanifold $\varphi(M)$ in N. When $\nabla d\varphi = 0$ on M, geodesics initially tangent to $\varphi(M)$ remain so and φ or $\varphi(M)$ is said to be <u>totally geodesic</u>. The <u>mean curvature</u> of φ is $\mu = (\dim M)^{-1} \mathrm{tr}(\nabla d\varphi)$, where

$$\text{(2.1)} \qquad \mathrm{tr}(\nabla d\varphi)_x = \sum_j (\nabla_{X_j} d\varphi)(X_j),$$

$\{X_j\}$ being an orthonormal basis of $T_x M$. Finally φ is said to have <u>constant mean curvature</u> if $\nabla^{\perp}\mu = 0$.

In general a mapping $\varphi : M \to N$ between Riemannian manifolds is said to be <u>harmonic</u> if $\mathrm{tr}(\nabla d\varphi) = 0$ [ESa]. Here, as in (2.1), the trace is taken with respect to the metric g. Using local coordinates $\varphi^1, \ldots, \varphi^n$ on N (by abuse of notation) and $x^1, \ldots, x^m$ on M, together with their associated Christoffel symbols,

$$\text{(2.2)} \qquad \mathrm{tr}(\nabla d\varphi) = g^{jk}\left[\frac{\partial^2 \varphi^r}{\partial x^j \partial x^k} - (\Gamma^M)^a_{jk} \frac{\partial \varphi^r}{\partial x^a} + (\Gamma^N)^r_{ab} \frac{\partial \varphi^a}{\partial x^j} \frac{\partial \varphi^b}{\partial x^k}\right] \frac{\partial}{\partial \varphi^r}.$$

Observe that the highest order term is just the linear Laplacian operator. Furthermore as far as N is concerned, only the connection plays a crucial role, and not the metric h.

One important aspect of harmonic maps that we do not pursue in the present work is the fact that they arise as solutions of a variational problem. For example if M is compact and φ is a Riemannian immersio it is well-known that the submanifold $\varphi(M)$ is minimal in the sense that its volume is critical iff $\mu = 0$, i.e. φ is harmonic. On the hand an arbitrary map φ is harmonic iff it corresponds to a critical point of the energy functional

$$E(\varphi) = \int_M e(\varphi) dV^M ,$$

where

$$e(\varphi) = \frac{1}{2} \|d\varphi\|^2 = \frac{1}{2} g^{jk} \frac{\partial \varphi^r}{\partial x^j} \frac{\partial \varphi^s}{\partial x^k} h_{rs}$$

By considering second variations of $E(\varphi)$ one can associate an index to a harmonic map. For more details refer to [ESa; EL].

When M and N are almost Hermitian manifolds, one can ask under what conditions holomorphic maps are necessarily harmonic. This is certainly the case if both M and N are Kähler, for on a Kähler manifold there exist complex coordinates for which the Christoffel symbols Γ^i_{jk} vanish at an assigned point. Choosing such coordinates at $x \in M$ and at $\varphi(x) \in N$ for (2.2) reduces $\mathrm{tr}(\nabla d\varphi)_x$ to the usual Laplacian, and one can argue as in the introduction. However the following more general result was proved by Lichnerowicz [Li, section 16]:

<u>Proposition 2.1</u> Let M,N be almost Hermitian manifolds with M cosymplectic and N (1,2)-symplectic. Then any holomorphic mapping $\varphi : M \to N$ is harmonic.

<u>Proof</u> Most of the work has already been done in section 1. Recall that cosymplectic means $\mathrm{tr}(D_2 F^M) = 0$, where F^M is the fundamental 2-vector of M. Let $\{\alpha_j\}$ be any local unitary basis of $T^{1,0}M$. By (1.4), $(\nabla_{\bar{\alpha}_j} \alpha_j)^{0,1} = \eta_{\bar{\alpha}_j} \alpha_j$, so in analogy to lemma 1.2 we have

$$\mathrm{tr}(D_2 F^M) = 0 \iff \sum_j \nabla_{\bar{\alpha}_j} \alpha_j \in T^{1,0}M.$$

Using $D_2 F^N = 0$ and lemma 1.2,

$$\mathrm{tr}(\nabla d\varphi) = \sum_j (\nabla_{\bar{\alpha}_j} d\varphi)(\alpha_j) = \sum_j \nabla_{\varphi_* \bar{\alpha}_j} (\varphi_* \alpha_j) - \varphi_* (\sum_j \nabla_{\bar{\alpha}_j} \alpha_j)$$

belongs to $T^{1,0}N$. However the left hand member is real, and therefore vanishes. ■

As a corollary, any holomorphic map φ between two (1,2)-symplectic manifolds is harmonic; in this case $\nabla d\varphi$ actually has no (1,1)-component. Such maps are discussed in [Ra, section 2]. A holomorphic submanifold of a (1,2)-symplectic manifold is itself (1,2)-symplectic with respect to the induced metric, and is therefore a minimal submanifolds [Gr_1].

It will become clear in the next section that the hypotheses of proposition 2.1 cannot be weakened (see [EL, section 9] for a counter-examples due to Gray). Lichnerowicz also proves in [Li] that holomorphic maps between almost Hermitian manifolds that are symplectic correspond to minima of the energy functional. Now [ES_2] show no such result true under the weaker assumptions of proposition 2.1, so the full generality is important.

Riemann surfaces

When M has only two real dimensions, there exists a concise formalism to compute $tr(\nabla d\varphi)$ for a mapping $\varphi . M \to N$. The point is that a metric and an orientation on M determine a unique Kähler structure, and M can be treated as a Riemann surface. The corresponding almost complex structure J^M is well-defined by the requirement that $J^M e_1 = e_2$ for any local oriented orthonormal frame $\{e_1, e_2\}$, and must be integrable. If $z = x + iy$ is a complex coordinate, the complex line bundles $T^{1,0}M$, $T^{0,1}M$ are spanned locally by

$$\frac{\partial}{\partial z} = \frac{1}{2}\left(\frac{\partial}{\partial x} - i\,\frac{\partial}{\partial y}\right), \qquad \frac{\partial}{\partial \bar{z}} = \frac{1}{2}\left(\frac{\partial}{\partial x} + i\,\frac{\partial}{\partial y}\right)$$

respectively. Hence $g(\partial/\partial z, \partial/\partial z) = 0$, which implies that $\partial/\partial x$, $\partial/\partial y$ are orthogonal with equal norms. The complex structure of M is equivalent to its oriented conformal structure.

Given a map $\varphi : M \to N$, set $\delta\varphi = \varphi_*(\partial/\partial z)$, $\bar{\delta}\varphi = \varphi_*(\partial/\partial\bar{z})$; by identifying the fibres of TN and $\varphi^{-1}TN$, we can regard these as sections of $\varphi^{-1}TN$. More generally, in order to take iterated derivatives, let $\delta, \bar{\delta}$ denote $\nabla_{\partial/\partial z}$, $\nabla_{\partial/\partial\bar{z}}$ respectively. Then $\mathrm{tr}(\nabla d\varphi)$ is proportional to

$$(\nabla_{\partial/\partial\bar{z}} d\varphi)(\frac{\partial}{\partial z}) = \nabla_{\partial/\partial\bar{z}}(\varphi_* \frac{\partial}{\partial z}) - \varphi_*(\nabla_{\partial/\partial\bar{z}} \frac{\partial}{\partial z}) = \bar{\delta}\delta\varphi.$$

In particular $\bar{\delta}\delta\varphi = \delta\bar{\delta}\varphi$ is real. A key observation is that whether or not φ is harmonic depends only upon the conformal class of M, so in future it makes sense to discuss maps $\varphi : M \to N$ from a Riemann surface into a Riemannian manifold. Such a map is said to be <u>conformal</u> if $h(\delta\varphi, \delta\varphi) = 0$ for any complex coordinate on M; this means that the induced metric φ^*h (where non-degenerate) is compatible with the conformal structure of M.

Properties of a conformal map $\varphi : M \to N$ may be characterized using the induced metric on the Riemann surface M relative to which the second fundamental form $\nabla^{\perp} d\varphi$ is determined by $(\delta^2\varphi)^{\perp}$ and $\bar{\delta}\delta\varphi$ (which is already normal). For example a conformal harmonic map φ has $\mu = 0$ relative to φ^*h, and is therefore minimal [ESa]. On the other hand

$$h(\delta^2\varphi, \delta\varphi) = \frac{1}{2}\delta h(\delta\varphi, \delta\varphi) = 0,$$

and $(\delta^2\varphi)^{\perp} = 0$ iff $\delta^2\varphi$ is proportional to $\delta\varphi$. If this holds at all points of M, φ is <u>totally umbilic</u>, a property depending only upon the conformal classes of M <u>and</u> N. When $N = \mathbb{R}^n$ and φ is totally umbilic, $\varphi(M)$ must lie in a plane or a sphere S^2 [C]; by conformal invariance the second conclusion also holds when $N = S^n$. Finally we say that φ is <u>totally geodesic</u> if it is both harmonic and totally umbilic, so that $\nabla^{\perp} d\varphi = 0$.

Any complex vector bundle E with a connection ∇ over a Riemann surface M has a canonical complex analytic structure [KM; AHS]. The total space of E is made into a complex manifold by using ∇ to "add" the complex structures of fibre and base. The zero section is then a holomorphic submanifold of the total space, so its normal bundle, naturally isomorphic to E, acquires a complex analytic structure. Complex analytic sections s of E are characterized by the equation $\nabla_{\partial/\partial\bar{z}} s = 0$. Applying this construction to the pullback $(\varphi^{-1}TN)^{\mathbb{C}}$ gives immediately

<u>Proposition 2.2</u> A map $\varphi : M \to N$ is harmonic iff $\delta\varphi$ is a local complex analytic section of $(\varphi^{-1}TN)^{\mathbb{C}}$.

This implies that the zeros of the differential φ_* are isolated, provided that φ is not constant. Moreover $[\delta\varphi]$ defines a global complex analytic section of the projective bundle $P((\varphi^{-1}TN)^{\mathbb{C}})$. This fact will be essential for future considerations, in which isolated zeros of φ_* do not require special attention. Singularities of harmonic maps between surfaces are discussed in [Wd]. The singularities of a conformal harmonic map $\varphi : M \to N$ from a surface into a Riemannian manifold are all branch points, i.e. in suitable coordinates

$$\varphi(z) = (\mathrm{Re}\, z^k, \mathrm{Im}\, z^k, 0, \ldots, 0) + \text{higher order terms}$$

[GOR]. From above such a map may then be considered to be the same as a <u>branched</u> <u>minimal</u> <u>immersion</u>, or <u>minimal</u> <u>surface</u> for short.

If M is a Riemann surface and N an almost Hermitian manifold, one can further decompose the differential of a map $\varphi : M \to N$. Choosing a complex coordinate on M, write

$$(2.3) \qquad \delta\varphi = \alpha + \bar{\beta}, \quad \alpha, \beta \in T^{1,0}N$$

(in alternative notation of [EW, Ra], $\alpha = D'\varphi = \nabla'\varphi$, $\beta = D''\varphi = \nabla''\varphi$). Thus φ is conformal iff $h(\alpha, \bar{\beta}) = 0$. If N is Kähler, φ is harmonic iff $\bar{\delta}\alpha = 0$, or equivalently iff $\delta\beta = 0$. In this case we shall have occasion to use iterated derivatives $\delta^r\alpha$, $\bar{\delta}^s\beta$, all of which belong to $T^{1,0}N$.

3. BUNDLES OF COMPLEX STRUCTURES

Let N be an oriented Riemannian manifold with even dimension 2n and metric h. We do not suppose that N is almost Hermitian, but at least the parity of the dimension ensures that almost complex structures exist locally. To study these, we introduce the bundle S whose fibre S_x over $x \in N$ consists of all the complex structures on the vector space T_xN compatible with the metric and orientation.

$$S_x = \{ J \in \mathrm{End}\, T_xN : J^2 = -1,\ h(JX,JY) = h(X,Y),\ J \geqslant 0 \}.$$

The expression $J \geqslant 0$ means that $X_1 \wedge JX_1 \wedge \ldots \wedge X_n \wedge JX_n$ is a non-negative multiple of the volume element for any vectors $X_1, \ldots, X_n$.

An oriented orthonormal basis of T_xN determines an isomorphism $T_xN \cong \mathbb{R}^{2n}$, and the action of $SO(2n)$ on $\mathbb{R}^{2n}$ then gives $S_x \cong SO(2n)/U(n)$. In other words S is the bundle with this homogeneous space as fibre associated to the principal $SO(2n)$-bundle of oriented orthonormal frames. For $n = 1$, $SO(2)/U(1)$ is just a point, and studying S amounts to considering N as a Riemann surface. In this section we shall see that S is also an almost complex manifold in higher dimensions. For $n = 2,3$ the fibres

$$\frac{SO(4)}{U(2)} \cong \frac{SU(2)}{U(1)}, \qquad \frac{SO(6)}{U(3)} \cong \frac{SU(4)}{S(U(1)\times U(3))}$$

are the projective spaces $\mathbb{C}P^1$, $\mathbb{C}P^3$ respectively, whereas by triality

$$\frac{SO(8)}{U(4)} \cong \frac{SO(8)}{SO(6)\times SO(2)}$$

is a complex quadric.

Because S is associated to the Riemannian structure of N, the Levi-Civita connection ∇ gives rise to a splitting

$$TS = H \oplus V \tag{3.1}$$

of the tangent bundle of S into horizontal and vertical components. The bundle V consists of all vectors tangent to the fibres, whereas H is best defined as follows. Associating to any $J \in S_x$ its fundamental 2-vector $F \in \Lambda^2 T_x N$ gives an embedding

$$i : S \hookrightarrow \Lambda^2 TN. \tag{3.2}$$

Fix $J \in S_x$, and extend it to a local section in such a way that the extended 2-vector satisfies $\nabla F|_x = 0$. Then the tangent space to this section at J is independent of the extension and equals the fibre of H.

Let $\pi : S \to N$ be the projection, and consider the pullback $\pi^{-1}TN$ whose fibre at $J \in S_x$ is naturally isomorphic to $T_x N$. Since J is an almost complex structure on $T_x N$, there is a decomposition

$$(\pi^{-1}TN)^{\mathbb{C}} = T^{1,0} \oplus T^{1,0}.$$

On the other hand π_* obviously induces an isomorphism $H \xrightarrow{\cong} \pi^{-1}TN$ whose inverse we denote by $\sigma \mapsto \sigma^h$. Hence

$$H^{\mathbb{C}} = (T^{1,0})^h \oplus (T^{0,1})^h, \tag{3.3}$$

which defines an almost complex structure J^h on the fibres of the real vector bundle H.

Fix $J \in S_x$ so that we can write

$$\Lambda^2 T_x N = T_J^{1,1} \oplus (T_J^{2,0} \oplus T_J^{0,2}).$$

Regarding the right-hand side as the sum of two real $U(n)$-modules, this is really the Lie algebra equation

$$so(2n) = u(n) \oplus m \tag{3.4}$$

for the homogeneous space $SO(2n)/U(n)$. Thus $T_J^{2,0} \oplus T_J^{0,2}$ is naturally isomorphic to the tangent space $T_J(S_x)$. The fact that $\Lambda^2(T_J^{2,0})$ and $T_J^{2,0} \otimes T_J^{0,2}$ contain no $U(n)$-submodules isomorphic to $T_J^{2,0}$ or $T_J^{0,2}$ implies that $[\mathfrak{m}, \mathfrak{m}] \subset \mathfrak{u}(n)$, and $SO(2n)/U(n)$ is a Hermitian symmetric space.

More invariantly, varying $J \in S$ gives

$$(\pi^{-1}\Lambda^2 TN)^{\mathbb{C}} = T^{1,1} \oplus (T^{2,0} \oplus T^{0,2}).$$

Now (3.2) induces a monomorphism $i_*: V \to \pi^{-1}\Lambda^2 TN$, and it is clear that

$$(i_* V)^{\mathbb{C}} = T^{2,0} \oplus T^{0,2}.$$

It is also convenient to consider the orthogonal projection

$$\pi^{-1}\Lambda^2 TN \to V$$

which we denote by $\sigma \mapsto \sigma^v$; its kernel is $T^{1,1}$, and

$$V^{\mathbb{C}} = (T^{2,0})^v \oplus (T^{0,2})^v \tag{3.5}$$

Define an almost complex structure J^v on the fibres of V by decreeing that J^v acts as i on $(T^{2,0})^v$ and $-i$ on $(T^{0,2})^v$. This seems the natural choice, and with this convention $(T^{2,0})^v$ is the bundle of (1,0)-vectors tangent to the complex fibres.

Combining (3.1), (3.3), (3.5) now yields

<u>Proposition 3.1</u> S has two distinct almost complex structures defined by $J_1 = J^h \oplus J^v$, $J_2 = J^h \oplus (-J^v)$.

The corresponding bundles of (1,0)-vectors are

$$T^{1,0}(S,J_1) = (T^{1,0})^h \oplus (T^{2,0})^v$$

(3.6)

$$T^{1,0}(S,J_2) = (T^{1,0})^h \oplus (T^{0,2})^v.$$

Switching between J_1 and J_2 is equivalent to reversing sign on the fibres, which in both cases are holomorphic submanifolds of S in the sense that their tangent spaces are stable under J_1 and J_2. On the other hand one can ask whether there are sections which are holomorphic submanifolds of S. Suppose that $f : U \to S$ is a section over some open set U of N with associated tensors $J \in \Gamma(U, \text{End}\, TN)$ and $F = i(J) \in \Gamma(U, \Lambda^2 TN)$. Then the tangent spaces of the submanifold $f(U)$ are stable under J_a iff the mapping $f : (U,J) \to (S,J_a)$ is holomorphic; in this case we shall say that f is J_a-holomorphic.

Proposition 3.2 f is J_a-holomorphic iff $D_a F = 0$.

Proof Fix $x \in N$ and let $y = f(x) \in S$. If $X \in T_x N$, then by lemma 1.1, $\nabla_X F \in T^{2,0}U \oplus T^{0,2}U$, the type decompositions being relative to J. Take a local basis $\{e_1, \ldots, e_{2n}\}$ of TN with $\nabla e_i|_x = 0$; the corresponding local trivialization of S is consistent with (3.1) at x. Writing $F = F^{ij} e_i \wedge e_j$, the vertical component of $f_* X$ at y is represented by

$$(XF^{ij}) e_i \wedge e_j = \nabla_X F|_x,$$

so more precisely,

$$f_* X = X^h + (\nabla_X F)^v. \tag{3.7}$$

Now f is J_a-holomorphic iff $f_* X \in T^{1,0}(S,J_a)$ for all $X \in T^{1,0}U$, but

$$\nabla_X F = \langle D_1 F, X \rangle + \langle D_2 F, X \rangle,$$

and $\langle D_1F,X\rangle \in T^{0,2}U$, $\langle D_2F,X\rangle \in T^{2,0}U$ for $X \in T^{1,0}U$. The proposition follows from (3.6). ■

Integrability properties

Propositions 1.3, 3.2 suggest that J_1 is more connected with complex structures, so let us suppose that the almost complex structure J_1 on S is integrable. In this case every point of S must belong to a J_1-holomorphic section $f : U \to S$. Then by lemma 1.2, the bundle $T^{1,0} = T^{1,0}U$ of (1,0)-vectors relative to f is preserved by ∇_X whenever $X \in T^{1,0}$.

Let R denote the Riemannian curvature tensor of N and take $X,Y,Z \in \Gamma(U;T^{1,0})$; then $[X,Y] \in T^{1,0}$ and

$$R(X,Y)Z = ([\nabla_X,\nabla_Y] - \nabla_{[X,Y]})Z \in T^{1,0}.$$

This means that $R(X,Y)$ has no component in the subbundle of $(\mathrm{End}\,TN)^{\mathbb{C}}$ isomorphic to $T^{0,2}$ (cf. (1.2)), whence R itself has no component in $T^{0,2} \otimes T^{0,2}$. If we fix $x \in N$, the type decompositions are all relative to the almost complex structure $J = f(x)$, so varying J over the fibre S_x gives

(3.8) $\quad R$ has no component in $W = \sum_{J\in S_x} T_J^{0,2} \otimes T_J^{0,2}$.

The space $W \subset (\Lambda^2 T_xN \otimes \Lambda^2 T_xN)^{\mathbb{C}}$ must be associated to some $SO(2n)$-module. Then since any tensor in W has zero Ricci contraction, condition (3.8) involves at most the Weyl conformal curvature of N. When $n \geqslant 3$ the latter is irreducible, so (3.8) is equivalent to N being conformally flat.

A special case has to be made for $\dim N = 4$, for then

$$\sum_{J\in S_x} T_J^{0,2}$$

is a proper subspaces of $(\Lambda^2 T_x N)^{\mathbb{C}}$ and coincides with the (complexification of) the +1 eigenspace Λ^2_+ of the $*$ operator. Therefore (3.8) is equivalent to the vanishing of that half of the Weyl tensor lying in $\Lambda^2_+ \otimes \Lambda^2_+$; when this occurs N is said to be <u>anti-self-dual</u>. If instead the Weyl tensor belongs to $\Lambda^2_+ \otimes \Lambda^2_+$, N is said to be <u>self-dual</u>. The special role played by the orientation in 4 dimensions is examined in the final section.

Conversely if (3.8) holds, J_1 is integrable. This is because if X,Y are vector fields on N, $R(X,Y)^v$ equals the vertical component of the Lie bracket $[X^h, Y^h]$. Using this fact one can show that $T^{1,0}(S,J_1)$ is closed under Lie bracket; see [BO; Sk; AHS] for more details. To summarize.

<u>Theorem 3.3</u> (S,J_1) is a complex manifold iff N^{2n} is conformally flat $(n \geq 3)$ or anti-self-dual $(n = 2)$.

Actually the construction of J_1 uses only the conformal structure of N, so any notion related to J_1 is conformally invariant. Based on ideas from Penrose's twistor programme, the complex manifold (S, J_1) was used in [AHS] to covert solutions of the Yang-Mills equations into holomorphic vector bundles. Accordingly we shall often refer to S as a <u>twistor space</u> or <u>twistor bundle</u> for N.

<u>Proposition 3.4</u> (S,J_2) is never integrable.

<u>Proof</u> Let $f : U \to S$ be a local section such that f(U) is a holomorphic submanifold relative to J_2 for which the induced almost complex structure J on U is integrable. If F is the corresponding fundamental 2-vector on U, by proposition 1.3 we have both $D_2F = 0$ and $D_1F = 0$. This forces f(U) to be horizontal, but if (S,J_2) were a complex manifold, there would exist many more such f. ■

The nonintegrability of J_2 is inherent in its definition, as is the next result which is a universal version of proposition 2.1.

<u>Theorem 3.5</u> Let M be a cosymplectic almost Hermitian manifold, and N a Riemannian manifold with twistor bundle $\pi : S \to N$. Then if $\psi : M \to S$ is J_2-holomorphic, $\pi \circ \psi$ is harmonic.

<u>Proof</u> Let $\varphi = \pi \circ \psi$, so that ψ may be regarded as a section over M of the pullback $\varphi^{-1}S \subset \varphi^{-1}\Lambda^2 TN$. Thus ψ defines an almost complex structure on the fibres of $\varphi^{-1}TN$ giving rise to a decomposition

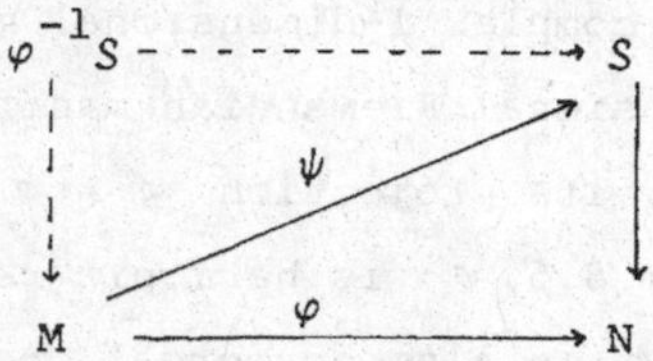

$$(3.9) \qquad (\varphi^{-1}TN)^{\mathbb{C}} = T^{1,0}N \oplus T^{0,1}N,$$

as well as a fundamental 2-vector $F \in \Gamma(M, \varphi^{-1}\Lambda^2 TN)$. In analogy with (1.6) we can write $\nabla F = D_1 F + D_2 F$, where now

$$D_1 F \in (T^{0,1}M \otimes T^{0,2}N) \oplus (T^{1,0}M \otimes T^{2,0}N)$$

$$D_2 F \in (T^{1,0}M \otimes T^{0,2}N) \oplus (T^{0,1}M \otimes T^{2,0}N).$$

By hypothesis, if $X \in T^{1,0}M$,

$$\psi_* X = (\varphi_* X)^h + (\nabla_X F)^v$$

(cf. (3.7)) belongs to $T^{1,0}(S, J_2)$, so $\varphi_* X \in T^{1,0}N$ and $\nabla_X F \in T^{0,2}N$. This means that $\varphi_* : TM \to \varphi^{-1}TN$ commutes with the respective almost complex structures, and these have the properties $\mathrm{tr}(D_2 F^M) = 0$, $D_2 F = 0$. This is exactly the situation of proposition 2.1, but with F^N replaced by F, so the proof there goes through. ■

4. MAPS OF SURFACES

The most obvious application of theorem 3.5 is to take M to be a Riemann surface; in this case there is no local obstruction to the existence of J_2-holomorphic maps $\psi : M \to S$. From now on we shall concentrate on this situation, so M will denote a Riemann surface, i.e. a connected complex 1-dimensional manifold, and as before $N = N^{2n}$ an even-dimensional Riemannian manifold. Given a J_2-holomorphic map $\psi : M \to S$, its projection $\varphi = \pi \circ \psi$ is a minimal surface in N. For by theorem 3.5, φ is harmonic relative to any compatible metric on M. However φ is also conformal, because for any complex coordinate on M, $\delta\varphi$ is a $(1,0)$ vector in TN relative to some almost Hermitian structure, so $h(\delta\varphi, \delta\varphi) = 0$.

Conversely one may ask whether an arbitrary minimal surface in N arises from a J_2-holomorphic curve in S. As a prerequisite to answering this question, we consider for any conformal immersion $\varphi : M \to N$ the subbundle $S'(\varphi)$ of $\varphi^{-1}S$ whose fibre over $x \in M$ is

$$S'(\varphi)_x = \{J \in S_{\varphi(x)} : J\,\delta\varphi(x) = i\,\delta\varphi(x)\}.$$

In other words $S'(\varphi)$ consists of all the almost complex structures which render φ holomorphic. Any $J \in S'(\varphi)_x$ induces an almost Hermitian structure on the orthogonal complement of $\varphi_*(T_xM)$ in $T_{\varphi(x)}N$, so $S'(\varphi)$ has fibre $SO(2n-2)/U(n-1)$. It is natural to think of $S'(\varphi)$ as a submanifold of S, but to be more precise we consider the natural map $\imath : S'(\varphi) \to S$ whose differential $\imath_*$ is everywhere injective. Thus we shall say that $S'(\varphi)$ is J_a-holomorphic if each tangent space $\imath_*(T_xS'(\varphi))$ is stable under J_a; in this case $S'(\varphi)$ is itself an almost complex manifold.

<u>Theorem 4.1</u> A conformal immersion $\varphi : M \to N^{2n}$ is harmonic iff $S'(\varphi)$ is J_2-holomorphic.

Proof Given the conformal immersion φ, let $W \subset \varphi^{-1}TN$ denote the normal bundle of M, so that there is an orthogonal sum

$$\varphi^{-1}TN \cong TM \oplus W.$$

If $p : \varphi^{-1}TN \to W$ denotes the projection, W has an induced connection $\overline{\nabla} = p \circ \nabla$. The complexification $W^{\mathbb{C}}$ then acquires a complex analytic structure with analytic sections s satisfying $\overline{\nabla}_{\partial/\partial \bar z}\, s = 0$ (see the remarks preceding proposition 2.2). Since the induced metric $h \in \Gamma(M, S^2 W^{\mathbb{C}})$ satisfies $\overline{\nabla} h = 0$, there is an associated complex analytic principal $SO(2n-2,\mathbb{C})$-bundle P. If C is the complex form of $SO(2n-2)/U(n-1)$ consisting of oriented totally isotropic subspaces of $W^{\mathbb{C}}$, then

$$S'(\varphi) \cong P \times_{SO(2n-2,\mathbb{C})} C.$$

Therefore $S'(\varphi)$ also acquires a complex analytic structure.

A section $\psi : U \to S'(\varphi)$, $U \subset M$, defines an isotropic rank $n-1$ subbundle $W^{1,0} \subset W^{\mathbb{C}}$. The image $\imath \circ \psi$ in S is the almost complex structure given by

$$T^{1,0}N = \mathbb{C}\delta\varphi \oplus W^{1,0}$$

in the notation of (3.9). Now ψ is analytic iff

$$\overline{\nabla}_{\partial/\partial \bar z} W^{1,0} \subset W^{1,0}. \tag{4.1}$$

On the other hand $\imath \circ \psi$ is J_2-holomorphic iff $D_2 F = 0$, where $F \in \Gamma(M, \varphi^{-1}\Lambda^2 TN)$ is the corresponding fundamental 2-vector (see the proof of theorem 3.5). By lemma 1.2 this is the case iff

$$\bar\delta(T^{1,0}N) \subset T^{1,0}N. \tag{4.2}$$

If φ is harmonic, then $\bar\delta\delta\varphi = 0$ and if s is a section of $W^{1,0}$, $h(\bar\delta s, \delta\varphi) = -h(s, \bar\delta\delta\varphi) = 0$. Thus (4.1) implies (4.2), and the complex analytic structure of $S'(\varphi)$ is the restriction of J_2. Conversely suppose that the tangent spaces of $S'(\varphi)$ are stable under J_2. Then given $m \in M$, there exists a section $\psi : U \to S'(\varphi)$, $m \in U \subset M$, such that (4.2) holds at m. But then $\bar{\delta}\delta\varphi \in T^{1,0}N$ and reality forces $\bar{\delta}\delta\varphi$ to vanish at m. Therefore φ is harmonic and once again $(S'(\varphi), J_2)$ is complex analytic. ∎

Suppose that $\varphi : M \to N$ is conformal and harmonic; as explained in section 2, the isolated zeros of φ_* do not affect the above construction. Hence $S'(\varphi)$ is a well-defined complex analytic bundle, any local analytic section of which is a J_2-holomorphic curve in S. Combining this with theorem 3.5 yields

Corollary 4.2 A map $\varphi : M \to N^{2n}$ is conformal and harmonic iff it is locally the projection of a J_2-holomorphic curve in S.

When $n = 1$, $S'(\varphi)$ is just a copy of M & a conformal harmonic φ has a unique global J_2-holomorphic lift; this case will be considered in more detail in section 8. Actually Rawnsley shows in [Ra, section 9] that the qualification "locally" in corollary 4.2 can always be omitted if M is compact. This is easy to see when $n = 3$ or 4, because $S'(\varphi)$ has a projective space as fibre and is therefore the projective bundle associated to some complex analytic vector bundle. Any meromorphic section of the latter, of which there are infinitely many, provides a J_2-holomorphic lift of φ.

A result of Nijenhuis & Woolf [NW; theorem III] asserts that for any almost complex manifold S with $X \in T_x^{1,0}S$, there exists a holomorphic curve $\psi : U \to S$, $0 \in U \subset \mathbb{C}$, with $\delta\psi = X$. The problem is first reduced in local coordinates to an equation of the form

$$\frac{\partial \psi^i}{\partial \bar{z}} + \sum_j a^i_j(\psi)\,\frac{\partial \overline{\psi^j}}{\partial \bar{z}} = 0, \qquad \frac{\partial \psi^i}{\partial z}(0) = X^i, \qquad a^i_j(0) = 0,$$

and it is shown that there are as many holomorphic curves through x in S as there would be in complex space of the same dimension. This fact together with the Riemannian submersion equations of O'Neill [O] can be used to construct minimal surfaces in N having preassigned jets.

We now turn attention to J_1.

<u>Corollary 4.3</u> Provided $n \geqslant 3$, a conformal immersion $\varphi : M \to N^{2n}$ is totally umbilic iff $S'(\varphi)$ is J_1-holomorphic.

<u>Proof</u> Modifying the final part of the proof of theorem 4.1 gives that $S'(\varphi)$ is J_1-holomorphic at some point or almost complex structure $J \in S'(\varphi)$ iff $\delta^2\varphi \in T_J^{1,0}N$, where $T_J^{1,0}N$ is the corresponding space of (1,0)-vectors. Hence $S'(\varphi)$ is J_1-holomorphic iff

$$\delta^2\varphi \in \bigcap_{J \in S'(\varphi)_x} T_J^{1,0}N, \quad \forall x \in M.$$

By definition the span $\mathbb{C}\,\delta\varphi$ is contained in this intersection, but for $n \geqslant 3$, it is easy to verify that they coincide. Therefore $S'(\varphi)$ is J_1-holomorphic iff $\delta^2\varphi$ is everywhere proportional to $\delta\varphi$. ■

The details of the above reflect an important distinction between J_1 and J_2. The projection of a J_1-holomorphic curve in S will not in general be totally umbilic; it will however have a weaker property called isotropy which we describe next. A map $\varphi : M \to N$ is said to be <u>real isotropic</u> (cf. [EW, definition 5.6; Ra, section 6]) if

$$h(\delta^r\varphi, \delta^s\varphi) = 0, \quad \forall r,s \geqslant 1 \tag{4.3}$$

(this is obviously independent of the complex coordinate used). The span of the $\delta^r\varphi$, $r \geqslant 1$, is then totally isotropic in $(TN)^{\mathbb{C}}$; the assumption that N has even dimension $2n$ ensures that this span can achieve a maximum of n complex dimensions. Putting $r = s = 1$ shows that isotropy generalizes conformality; note also that being real isotropic depends only on the conformal structures of M and N.

Proposition 4.4 The projection of a J_1-holomorphic curve in S is real isotropic.

Proof Let $\varphi = \pi \circ \psi$ be the projection of a J_1-holomorphic curve $\psi : M \to S$. In the notation of the proof of theorem 3.5,

$$\delta\psi = (\delta\varphi)^h + (\nabla_{\partial/\partial z} F)^v$$

belongs to $T^{1,0}(S, J_1)$ which implies that $\delta\varphi \in T^{1,0}N$ and $D_1 F = 0$. Using lemma 1.2 it follows by induction that

$$\delta^{r+1}\varphi = \nabla_{\partial/\partial z}(\delta^r\varphi)$$

belongs to $T^{1,0}N$ for all $r \geqslant 1$, so (4.3) holds. ∎

It follows from propositions 3.2 and 4.4 (or directly from lemma 1.2) that any holomorphic mapping from a Riemann surface into a Hermitian manifold is real isotropic; this is an analogue of proposition 2.1.

Maps into spheres

Let S^k denote the sphere of unit vectors in $\mathbb{R}^{k+1}$, with the induced Euclidean metric h. It is known that any harmonic map $\varphi : S^2 \to S^k$ must be isotropic [Ca_2; Ch]. This uses the fact that the Riemann surface S^2 has no holomorphic differentials, and that the Riemannian curvature tensor R of S^k is a multiple c of the identity. For instance

$$\bar{\delta} h(\delta\varphi, \delta\varphi) = 2h(\bar{\delta}\delta\varphi, \delta\varphi) = 0,$$

so the globally-defined form $h(\delta\varphi, \delta\varphi)dz^2$ on S^2 is holomorphic; thus $h(\delta\varphi, \delta\varphi) = 0$ and φ is conformal (true for any harmonic $\varphi : S^2 \to N$). To establish (4.3) it suffices to prove that $\bar{\delta}h(\delta^r\varphi, \delta^s\varphi) = 0$, but this follows by induction using

$$(4.4) \qquad (\delta\bar{\delta}-\bar{\delta}\delta)(X) = R(\delta\varphi,\bar{\delta}\varphi)(X)$$

$$= c(h(\delta\varphi,X)\bar{\delta}\varphi - h(\bar{\delta}\varphi,X)\delta\varphi).$$

Given a real isotropic harmonic map $\varphi: M \to S^k$, let n be the maximum dimension attained by the span of $\delta^r\varphi$, $r \geq 1$. Using (4.4) one can show that if $n < \frac{1}{2}k$, $\varphi(M)$ must be contained in some hypersphere, i.e. φ is not full. Therefore for a full real isotropic harmonic map of M into S^k, $k = 2n$ is necessarily even [Ca_2].

The action of $SO(2n+1)$ on $\mathbb{R}^{2n+1}$ gives

$$S^{2n} \cong \frac{SO(2n+1)}{SO(2n)}$$

$$\mathcal{S} = \mathcal{S}(S^{2n}) \cong \frac{SO(2n+1)}{U(n)}$$

By theorem 3.3, J_1 makes the latter into a complex manifold. In fact the inclusion $SO(2n+1) \subset SO(2n+2)$ induces an isomorphism

$$\frac{SO(2n+1)}{U(n)} \cong \frac{SO(2n+2)}{U(n+1)}$$

of $\mathcal{S}(S^{2n})$ with the fibre of $\mathcal{S}(N^{2n+2})$ in which J_1 coincides with J^v (see (3.5)). A map $\psi: M \to \mathcal{S}$ is said to be <u>horizontal</u> if each tangent space $\psi_*(T_xM)$ lies in the horizontal distribution H. For such a map, the adjective "holomorphic" needs no qualification since J_1 and J_2 coincide on H.

The next result is due to Calabi [Ca_3], and has been generalized by Burstall & Rawnsley [Ra, theorem 6.10]. It converts a harmonic map into a genuine holomorphic curve in a complex manifold, and a closely related construction due to Chern [Ch_2, B] produces a holomorphic curve in the complex projective space $\mathbb{C}P^{2n}$ (see the end of section 7).

Theorem 4.5 There is a 2 : 1 correspondence between full real isotropic harmonic maps $\varphi : M \to S^{2n}$ and full horizontal holomorphic curves $\psi : M \to S$.

Proof To make things easy, we define ψ to be full iff $\pi \circ \psi$ is full in S^{2n}. Given φ, it follows from proposition 2.2 that $\delta\varphi \wedge \ldots \wedge \delta^n\varphi$ is non-zero on a dense subset of M and that span $\{\delta\varphi, \ldots, \delta^n\varphi\}$ is contained in a unique complex rank n subbundle $T^{1,0}$ of $(\varphi^{-1}TS^{2n})^{\mathbb{C}}$. By replacing φ by $a \circ \varphi$ where $a : S^{2n} \to S^{2n}$ is the orientation-reversing antipodal map, we may assume that the corresponding almost complex structure J on $\varphi^{-1}TN$ satisfies $J \geqslant 0$. Thus J defines a mapping $\psi : M \to S$ which will be both J_1 and J_2 holomorphic iff $T^{1,0}$ is stable under both δ and $\bar{\delta}$ (see (4.2)). Isotropy implies that $\delta^r\varphi \in T^{1,0}$ for all r, whereas equation (4.4) and $\bar{\delta}\delta\varphi = 0$ imply that $h(\bar{\delta}\delta^r\varphi, \delta^s\varphi) = 0$ for all r,s, and so $\bar{\delta}\delta^r\varphi \in T^{1,0}$. Conversely one associates to a given ψ the two maps $\pi \circ \psi$ and $a \circ \pi \circ \psi$. ∎

Orientation was an important feature in the above proof. Given a full real isotropic harmonic map $\varphi : M \to S^{2n}$, define $T^{1,0}$ as before and choose $\tau \in \Gamma(M, T^{1,0})$ with $h(\delta^r\varphi, \bar{\tau}) = 0$, $1 \leqslant r \leqslant n-1$. Then using instead the almost complex structure J_- for which $\delta\varphi, \ldots, \delta^{n-1}\varphi, \bar{\tau}$ are (1,0) vectors, one can define a J_2-holomorphic curve

$$\psi_- : M \to SO(2n+1)/U(n)$$

which unlike $\psi = \psi_+$ is not J_1-holomorphic.

For completeness one should therefore consider over an oriented even-dimesional Riemannian manifold N, not only $S = S_+$, but also the bundle S_- of almost complex structures compatible with the metric and reversed orientation $(J \leqslant 0)$. This will be done covertly in the next section, and explicitly in section 8. A converse to proposition 4.4 now emerges: any real isotropic map $M \to N$ is locally the projection of a J_1-holomorphic curve in either S_+ or S_-.

5. SYMMETRIC SPACES

The bundle $\mathcal{S}$ with fibre $SO(2n)/U(n)$ is generally too large a manifold to work with, and unfortunately $(\mathcal{S}, J_1)$ is rarely integrable (theorem 3.3). However most of the results of the last section are applicable to holomorphic submanifolds of $\mathcal{S}$ which arise naturally when the base is a symmetric space.

Let $N = G/H$ be a Riemannian symmetric space of even dimension $2n$ with G acting almost effectively, and H connected. The choice of an orthonormal frame at the identity coset determines both an orientation for N, and a linear isotropy representation

$$\rho : H \longrightarrow SO(2n)$$

with discrete kernel. Here $SO(2n)$ is the special orthogonal group of a fixed vector space $\mathbb{R}^{2n}$, so let $U(n) \subset SO(2n)$ denote the subgroup fixing the standard almost complex structure on $\mathbb{R}^{2n}$, and Z the centre of $U(n)$. Putting $K = \{h \in H \mid \rho(h) \in U(n)\}$ gives an inclusion of homogeneous spaces

(5.1) $$H/K \hookrightarrow SO(2n)/U(n)$$

Then $\mathcal{T} = G/K$ is isomorphic to the bundle with fibre H/K associated to the principal bundle $G \longrightarrow N$, and there is a natural inclusion $\mathcal{T} \subset \mathcal{S}$.

<u>Theorem 5.1</u> If $Z \subset \rho(H)$, then $\mathcal{T}$ is a holomorphic submanifold of $\mathcal{S}$ for both J_1 and J_2, with $(\mathcal{T}, J_1)$ Hermitian and $(\mathcal{T}, J_2)$ (1,2)-symplectic for some G-invariant metric.

<u>Proof</u> For the convenience of the proof, suppose that G is centreless so that ρ is injective and we can identify $\rho(H)$ with H. Let g, h, k denote the complexifications of the Lie algebras of G, H, K.

There is a decomposition

$$\mathfrak{g} = \mathfrak{h} \oplus \mathfrak{m}, \ [\mathfrak{h},\mathfrak{m}] \subset \mathfrak{m}, \ [\mathfrak{m},\mathfrak{m}] \subset \mathfrak{h}.$$

in which $\mathfrak{m}$ can be regarded as the restriction to H of the standard representation of $SO(2n)$. Restricting further to $K \subset U(n)$ therefore gives

$$\mathfrak{m} = \mathfrak{m}^{1,0} \oplus \mathfrak{m}^{0,1}$$

in which $\mathfrak{m}^{1,0}, \mathfrak{m}^{0,1}$ are the eigenspaces of a non-trivial element of $Z \subset K = H \cap U(n)$. We shall say that a Z-module has weight r if the scalar matrix $e^{it}.1 \in U(n)$ acts as e^{rit}; then $\mathfrak{m}^{1,0}, \mathfrak{m}^{0,1}$ have weights $+1, -1$ respectively.

As a $U(n)$-module, the Lie algebra of $SO(2n)$ splits naturally as

$$\mathfrak{so}(2n) = \mathfrak{u}(n) \oplus (\mathfrak{m}^{2,0} \oplus \mathfrak{m}^{0,2})$$

where $\mathfrak{u}(n) \cong \mathfrak{m}^{1,1}$ (cf. (3.4)), and the summands on the right hand side have weights $0,2,-2$ respectively. Decomposing the submodule $\mathfrak{h}$ of $\mathfrak{so}(2n)$ according to weights,

$$\mathfrak{h} = \mathfrak{k} \oplus (\mathfrak{n}^{1,0} \oplus \mathfrak{n}^{0,1}),$$

where $\mathfrak{k} = \mathfrak{h} \cap \mathfrak{u}(n)$, $\mathfrak{n}^{1,0} = \mathfrak{h} \cap \mathfrak{m}^{2,0}$, $\mathfrak{n}^{0,1} = \mathfrak{h} \cap \mathfrak{m}^{0,2}$. Moreover if $[\cdot,\cdot]$ denotes Lie bracket in $\mathfrak{h}$ or $\mathfrak{so}(2n)$, $[\mathfrak{n}^{1,0},\mathfrak{n}^{1,0}]$ has weight 4 and so vanishes. In this way we obtain

$$[\mathfrak{n}^{1,0},\mathfrak{n}^{1,0}] = 0 = [\mathfrak{n}^{0,1},\mathfrak{n}^{0,1}], [\mathfrak{n}^{1,0},\mathfrak{n}^{0,1}] \subset \mathfrak{h}.$$

Consequently H/K is a Hermitian symmetric space, and the inclusion (5.1) is holomorphic.

The holonomy of N reduces to H, and so the Levi-Civita connection on S reduces to

$$T = G \times_H H/K \subset G \times_H {}^{SO(2n)}/_{U(n)} = S \,.$$

This means that at any $x \in T$, the horizontal subspace of $T_x S$ defined by (3.1) actually lies in $T_x T$. Then by (3.6), J_1 and J_2 are both homogeneous almost complex structures of T (see Borel & Hirzebruch [BH, chapter 4]). Indeed $g = k \oplus m \oplus n$, where we can identify

$$T_x^{1,0}(T, J_1) = m^{1,0} \oplus n^{1,0}$$

$$T_x^{1,0}(T, J_2) = m^{1,0} \oplus n^{0,1}$$

m being the horizontal space, and n the vertical.

Using weights,

$$[m^{1,0} \oplus n^{1,0}, m^{1,0} \oplus n^{1,0}] \subset n^{1,0},$$

which implies that the subbundle $T^{1,0}(T, J_1)$ is closed under Lie bracket. Therefore (T, J_1) is a complex manifold, and Hermitian relative to any G-invariant almost Hermitian metric. As for J_2, put

$$a^0 = k, \quad a^1 = m^{1,0} \oplus n^{0,1}, \quad a^2 = m^{0,1} \oplus n^{1,0},$$

and observe that

$$[a^i, a^j] \subset a^k, \quad k = i + j \pmod 3).$$

If ω denotes a primitive cube root of unity, the transformation P of g with eigenspaces a^0, a^1, a^2 and eigenvalues $1, \omega, \omega^2$ respectively is a Lie algebra automorphism satisfying $P^3 = 1$. Such automorphisms were studied by Wolf & Gray, and [WG, theorem 8.13] implies that any G-invariant almost Hermitian metric for J_2 is automatically (1,2)-symplectic. ■

Table to determine $T = G/K$, G compact

G	H	K	H/K
$SO(m+2n)$, $n \geq 2$	$SO(m) \times SO(2n)$	$SO(m) \times U(n)$	$SO(2n)/U(n)$
$SO(2n)$, $n \geq 3$	$U(n)$	$SO(2) \times U(n-1)$	$\mathbb{C}P^{n-1}$
$SU(m+n)$	$S(U(m) \times U(n))$	$S(U(m) \times U(k) \times U(n-k))$	$G_k(\mathbb{C}^n)$
$Sp(m+n)$	$Sp(m) \times Sp(n)$	$U(m) \times Sp(n)$	$Sp(n)/U(n)$
G_2	$SO(4)$	$U(2)$	$\mathbb{C}P^1$
F_4	$Sp(3)Sp(1)$	$Sp(3)U(1)$	$\mathbb{C}P^1$
	$Spin(9)$	$Spin(7)U(1)$	Q^7
E_6	$SU(6)Sp(1)$	$SU(6)U(1)$	$\mathbb{C}P^1$
		$S(U(5) \times U(1))Sp(1)$	$\mathbb{C}P^5$
	$SO(10)SO(2)$	$(SO(8) \times SO(2))SO(2)$	Q^8
E_7	$Spin(12)Sp(1)$	$Spin(12)U(1)$	$\mathbb{C}P^1$
		$(Spin(10)U(1))Sp(1)$	Q^{10}
	$SU(8)/\mathbb{Z}_2$	$S(U(7) \times U(1))/\mathbb{Z}_2$	$\mathbb{C}P^7$
E_8	$E_7Sp(1)$	$E_7U(1)$	$\mathbb{C}P^1$
	$Spin(16)/\mathbb{Z}_2$	$(Spin(14)U(1))/\mathbb{Z}_2$	Q^{14}

If G is a Lie group with an automorphism of order 3, and K is the subgroup of fixed points, then G/K is called a <u>3-symmetric space</u> [Gr_4]. Each point of G/K is the isolated point of a 3-fold symmetry. The classification of 3-symmetric spaces [WG, theorem 7.10] makes it easy to enumerate the spaces T that arise in theorem 5.1. Given $N = G/H$, one seeks a subgroup K of H for which H/K is Hermitian symmetric and G/K is 3-symmetric. The case $H = K$ is also admissible, and corresponds to $N = T$ being Hermitian symmetric. For G compact and $H \neq K$, the table provides a complete list of the possibilities, which are precisely those entries of [WG, table 1, section 6] having at least 4 invariant almost complex structures. In our notation juxtaposition AB of two groups means $(A \times B)/\mathbb{Z}_2$, and Q^k denotes

$$\frac{SO(k+2)}{SO(k) \times SO(2)} \cong \tilde{G}_2(\mathbb{R}^{k+2}) \tag{5.2}$$

which can be identified with the complex hyperquadric in $\mathbb{C}P^{k+1}$ [Ch_1].

The hypothesis $Z \subset \rho(H)$ means that relative to ρ, the Lie algebra $\mathfrak{h}$ of H contains the standard complex structure on $\mathbb{R}^{2n}$. In other words there exists $j \in \mathfrak{h}$ such that $(\mathrm{ad}_{\mathfrak{m}} j)^2 = -1$, and such an element is called a <u>twistor structure</u> by Bryant. In [Br_4] these twistor structures are classified directly, and it is shown that the corresponding T can be characterized as integrable J_1-holomorphic subbundles of S.

Theorem 5.1 is an example of a more general construction of twisto spaces described in [BO; OR], and which arose partly from consideration of the quaternionic case [S_3]. Starting from a 2n-dimensional manifold N with a G-structure, $G \subset GL(2n,\mathbb{R})$, one may consider an associated bundle whose fibre is some G-invariant complex submanifold of the complex manifold $GL(2n,\mathbb{R})/GL(n,\mathbb{C})$. A G-connection ∇ then equips the total space with an almost complex structure, whose integrability tensor may be expressed in terms of the torsion and curvature of ∇.

For the G-invariant metric alluded to in theorem 5.1, we may take any Riemannian metric of the form

$$k_t = tg + \pi^*h,$$

where g is a vertical metric induced from the Riemannian symmetric structure of the fibre H/K, h is the Riemannian metric on N, and t is a positive constant. The projection $G/K \rightarrow G/H$ is then a Riemannian submersion. There will exist a unique value of t for which k_t is induced from a bi-invariant metric on G, and the (1,2)-symplectic manifold (T, J_2, k_t) is then S^6-like, i.e. $D_2F \in T^{3,0} \oplus T^{0,3}$ [WG, theorem 8.13]. Incidentally, $S^6 = G_2/SU(3)$ itself is an example of a 3-symmetric space with irreducible isotropy.

With respect to the metrics k_t (any $t > 0$) and h, proposition 2.1 and theorems 3.5, 5.1 combine to give

Corollary 5.2 Any J_2-holomorphic curve $M \rightarrow T$ is itself a minimal surface, and projects to a minimal surface in N.

Of the two assertions of the corollary, the second is the most natural and can be applied to a wider class of homogeneous fibrations in which the base is symmetric. The important property of J_2 in this regard is the fact that the fibres H/K are J_2-holomorphic submanifolds and that the almost complex structure J_1 obtained by reversing the sign of J_2 vertically is integrable.

6. KÄHLER GEOMETRY

The table in the last section shows that the complex Grassmannian $G_m(\mathbb{C}^{m+n})$ admits a twistor bundle

$$\mathcal{T}_k = \frac{U(m+n)}{U(m) \times U(k) \times U(n-k)}, \quad 1 \leqslant k \leqslant n-1,$$

which is a flag manifold. One important observation is that, up to sign, J_2 is the only invariant almost complex structure on $\mathcal{T}_k$ which is not integrable. For the isotropy representation has three irreducible real components, so by results of Borel & Hirzebruch [BH] there are $2^3 = 8$ invariant almost complex structures, of which $3! = 6$ are integrable.

Taking $m = 1$, we obtain as base the complex projective space $\mathbb{C}P^n$, and $\mathcal{T}_k$ may be identified with the bundle $G_k(T^{1,0}\mathbb{C}P^n)$ whose fibre at x is the Grassmannian of complex k-dimensional subspaces of the holomorphic tangent space $T_x^{1,0}\mathbb{C}P^n$. If J denotes the standard complex structure of $\mathbb{C}P^n$, then an element $W \in (\mathcal{T}_k)_x$ also represents a real 2k-dimensional J-invariant subspace of the real tangent space $T_x\mathbb{C}P^n$. The corresponding almost complex structure on $T_x\mathbb{C}P^n$ defined by $W \in \mathcal{T}_k \subset \mathcal{S}$ is

$$J_W = \begin{cases} J & \text{on } W \\ -J & \text{on } W^{\perp}. \end{cases} \tag{6.1}$$

Now let N be any Kähler manifold. The prescription (6.1) can be used to realize the Grassmannian bundle $G_k(T^{1,0}N)$ as a subbundle of $\mathcal{S}$. The Kähler assumption then ensures that the Levi-Civita connection and the almost complex structures J_1, J_2 reduce from $\mathcal{S}$ to $G_k(T^{1,0}N)$. This construction was developed by O'Brian & Rawnsley in [OR]. Note that when N is $G_m(\mathbb{C}^{m+n})$ with $m,n \geqslant 2$, $G_k(T^{1,0}N)$ is no the same as $\mathcal{T}_k$

Let φ $M \longrightarrow N$ be a conformal immersion of a Riemann surface into a Kähler manifold, and set

$$\delta\varphi = \alpha + \overline{\beta}, \quad \alpha,\beta \in T^{1,0}N, \quad h(\alpha,\overline{\beta}) = 0$$

as in (2.3). The real 2-planes $i\alpha \wedge \overline{\alpha}$, $i\beta \wedge \overline{\beta}$ are mutually orthogonal, so there exists a real 2k-dimensional J-invariant subspace $W \subset T_xN$ with $i\alpha \wedge \overline{\alpha}$ in W and $i\beta \wedge \overline{\beta}$ in $W^{\perp}$. This means that

$$S'(\varphi) \cap \varphi^{-1}G_k(T^{1,0}N)$$

is nonempty, and by theorem 4 1 will be J_2-holomorphic iff φ is harmonic.

When $k = 1$ one is dealing with the projective holomorphic tangent bundle $G_1(T^{1,0}N) = P(T^{1,0}N)$ of N. In this case given a conformal harmonic φ, there exists a unique J_2-holomorphic map ψ. $M \to P(T^{1,0}N)$ with $\varphi = \pi \circ \psi$, provided φ itself is not antiholomorphic. The lift is simply the projective class $[\alpha]$ which is everywhere defined even if φ_* has isolated zeros, since α is a complex analytic section of $(\varphi^{-1}TN)^{\mathbb{C}}$. Conversely any J_2-holomorphic map ψ. $M \to P(T^{1,0}N)$ projects to a conformal harmonic map φ $M \to N$ whose differential may have isolated zeros In the special situation in which the (1,0)-component of the differential ψ_* is vertical, there may be many J_2-holomorphic maps projecting to the same antiholomorphic φ, but we can state

<u>Theorem 6.1</u> There is a generically 1 : 1 correspondence between minimal surfaces in a Kähler manifold N, and J_2-holomorphic curves in $P(T^{1,0}N)$.

The flag manifold $\mathcal{T}_1$ is the total space of a triple fibration

$$(6.2)\qquad \begin{array}{ccccc} & & \mathcal{T}_1 & & \\ & \swarrow^{\pi} & \downarrow^{\pi'} & \searrow^{\pi''} & \\ \mathbb{C}P^n & & \mathbb{C}P^n & & G_2(\mathbb{C}^{n+1}). \end{array}$$

Theorem 5.1 may be applied to each projection π, π', π'', but the corresponding nonintegrable almost complex structures J_2, J_2', J_2'' must all agree up to sign. By above, any conformal harmonic map $\varphi : M \to \mathbb{C}P^n$ which is not antiholomorphic defines a J_2-holomorphic map $\psi : M \to \mathcal{T}_1$. Defining the <u>transforms</u> of φ to be the other projections

$$\varphi' = \pi' \circ \psi, \quad \varphi'' = \pi'' \circ \psi,$$

corollary 5.2 gives

<u>Corallary 6.2</u> The transforms φ', φ'' of the minimal surface φ are themselves minimal surfaces.

Starting from φ, for any $k \geqslant 1$ one can form the k-fold transform $\varphi^{(k)} = ((\varphi')'..)'$. If φ is itself holomorphic (i.e. $\beta \equiv 0$), then $\varphi^{(n)}$ is antiholomorphic, and so after n steps the process has terminated. This is the essence of a classification theorem of Eells & Wood [EW, theorem 6.9] based on [DZ], which implies that any harmonic map from S^2 into $\mathbb{C}P^n$ equals an iterated transform $h^{(k)}$ of some holomorphic map $h : S^2 \to \mathbb{C}P^n$; see also [Bu,GS,Wo]. Inherent in this construction is the fact that for $n \geqslant 3$, a J_2-holomorphic curve in $\mathcal{T}_1$ can be "differentiated" to produce a new J_2-holomorphic curve in $\mathcal{T}_1$.

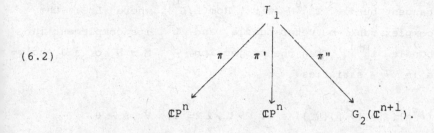

Given a Riemannian symmetric space N with twistor bundle $\mathcal{T}$ as in theorem 5.1, any J_1-holomorphic curve $\psi : M \to \mathcal{T}$ projects to a real isotropic map $\varphi : M \to N$ by proposition 4.4. However the assumption that ψ lies in the subbundle $\mathcal{T}$ of S makes φ isotropic in a more restrictive sense. As an example take $N = G_m(\mathbb{C}^{m+n})$, and identify the holomorphic tangent bundle $T^{1,0}N$ with $\mathrm{Hom}(L,L^{\perp})$, where L is the tautologous complex rank m vetor bundle, and $L^{\perp}$ its complement in the trivial bundle $\underline{\mathbb{C}}^{m+n}$. Then the projection $\varphi : M \to N$, of a J_1-holomorphic curve in T_k satisfies

$$(6.3) \qquad h((\delta^r \alpha)(\ell_1),(\delta^s \overline{\beta})(\ell_2)) = 0, \quad \forall \ell_1, \ell_2 \in L, \quad \forall r,s \geq 0.$$

In [ErW], property (6.3) is called <u>strong isotropy</u>; for complex projective space $m = 1$, it reduces to

$$(6.4) \qquad h(\delta^r \alpha, \delta^s \overline{\beta}) = 0 \quad \forall r,s \geq 0$$

which in [EW] is called <u>complex isotropy</u>. This second notion makes sense for maps to any Kähler manifold N; indeed it corresponds to the J_1-structure on the twistor space $G_k(T^{1,0}N)$.

Conversely, suppose that $\varphi : M \to G_m(\mathbb{C}^{m+n})$ is a strongly isotropic harmonic map which is full, i.e. not contained in some $G_m(\mathbb{C}^r)$, $r < m+n$. Then there exists k, $0 \leq k \leq n$, for which generically

$$\dim_{\mathbb{C}} \mathrm{span}\{(\delta^r \alpha)(\ell) \ . \ r \geq 0\} = k, \quad \forall \ell \in L;$$

in this case φ may be said to have type k. In analogy to theorem 4.5, one obtains a $1 : 1$ correspondence between strongly isotropic harmonic maps $\varphi : M \to G_m(\mathbb{C}^{m+n})$ of type k, and full horizontal holomorphic curves $\psi : M \to T_k$. This is essentially Erdem & Woods' [ErW, theorem 1.1].

<u>Quaternionic Kähler manifolds</u>

For each Lie group G in the table in section 5, there exists a subgroup K and a twistor space $T = G/K$ with fibre $H/K = \mathbb{C}P^1$. In this case the Riemannian symmetric space $N = G/H$ is called <u>quaternionic Kähler</u> because its linear holonomy factors through $Sp(n)Sp(1) \subset SO(4n)$

[W, S_1]. Notice that a quaternionic Kähler manifold is not necessarily Kähler in the ordinary sense; indeed there may be no almost complex structure which is globally defined. However each point $x \in N$ possesses a neighbourhood U on which there exist almost complex structures $I, J, K \in \Gamma(U, T)$ satisfying $IJ = -JI = K$; the fibre of T over x then parametrizes the 2-sphere

$$\{aI + bJ + cK : a^2 + b^2 + c^2 = 1\}$$

of complex structures on $T_x N$.

Any vector $X \in T_x N$ generates a quaternionic 1-dimensional subspace which is spanned by X, IX, JX, KX, and a mapping $\varphi : M \to N$ is said to be inclusive if each tangent space $\varphi_*(T_x M)$ is contained in a subspace of this type [ES_1]. For a conformal inclusive immersion φ, the fibre of

$$S'(\varphi) \cap \varphi^{-1} T$$

is a single point, and this time we obtain a $1:1$ correspondence between inclusive minimal surfaces in N and J_2-holomorphic curves in T that are not contained in a fibre. Corollary 5.2 then gives

Theorem 6.3 An inclusive minimal surface in a quaternionic Kähler symmetric space N lifts to a minimal surface in the twistor bundle T with fibre $\mathbb{C}P^1$.

Quaternionic projective space $\mathbb{H}P^n$ has a twistor bundle

$$(6.5) \qquad T = \frac{Sp(n+1)}{U(1) \times Sp(n)} \cong \mathbb{C}P^{2n+1}$$

which has been used by Jensen & Rigoli [JR] to study minimal surfaces in $\mathbb{H}P^n$. Theorem 6.3 and [EW, theorem 6.9] imply that any inclusive minimal surface $S^2 \to \mathbb{H}P^n$ is the projection of the transform $h^{(k)}$

of some holomorphic map $h : S^2 \to \mathbb{C}P^{2n+1}$. We illustrate this by considering some homogeneous maps mentioned in [G].

From the proof of theorem 5.1,

$$sp(n+1) = u(1) \oplus sp(n) \oplus (m^{1,0} \oplus m^{0,1}) \oplus (n^{1,0} \oplus n^{0,1}).$$

If $\zeta \cong \mathbb{C}$, $E \cong \mathbb{C}^{2n}$ denote the basic representations of the subgroups $U(1)$, $Sp(n)$ respectively in (6.5), then $sp(n+1) \cong S^2(\zeta \oplus \bar{\zeta} \oplus E)$ and we may take

$$m^{1,0} \cong \zeta \otimes E, \quad n^{1,0} \cong \zeta \otimes \zeta = \zeta^2.$$

Now take another copy of $U(1)$ with basic representation η. Relative to the above structure there are then homomorphisms $\Psi_r : U(1) \to Sp(n+1)$ defined by the equations

$$\zeta = \eta^r, \quad E \oplus \zeta \oplus \bar{\zeta} = S^{2n+1}(\eta \oplus \bar{\eta}),$$

where

$$S^{2n+1}(\eta \oplus \bar{\eta}) = \eta^{2n+1} \oplus \eta^{2n-1} \oplus \ldots \oplus \bar{\eta}^{(2n-1)} \oplus \bar{\eta}^{(2n+1)};$$

thus r is odd and $|r| \leqslant 2n+1$. Now the basic representation of $Sp(1)$ on $\mathbb{C}^2$ decomposes as $\eta \oplus \bar{\eta}$ relative to the standard inclusion $i : U(1) \subset Sp(1)$. Therefore by construction, Ψ_r factors through i to give a map

$$\psi_r : S^2 \cong \mathbb{C}P^1 \cong \frac{Sp(1)}{U(1)} \longrightarrow \frac{Sp(n+1)}{U(1) \times Sp(n)} \cong \mathbb{C}P^{2n+1}.$$

The Lie algebra equation for the domain $\mathbb{C}P^1$ is

$$sp(1) \cong u(1) \oplus (\eta^2 \oplus \bar{\eta}^2),$$

and properties of ψ_r can be ascertained by checking which of $m^{1,0}$, $m^{0,1}$, $n^{1,0}$, $n^{0,1}$ contain η^2. Decreeing $\eta^2 \cong T^{0,1}\mathbb{C}P^1$ gives

ψ_1	J_2-antiholomorphic, not horizontal
$\psi_3,\dots,\psi_{2n-1}$	horizontal, not holomorphic (nor anti)
ψ_{2n+1}	horizontal holomorphic.

with corresponding statements for ψ_{-r}.

Actually ψ_{2n+1} is essentially the Veronese curve in $\mathbb{C}P^{2n+1}$, and the other ψ_r are the transforms of ψ_{2n+1}, which are also conformal harmonic (corollary 6.2). Because the composition of a horizontal harmonic map with a Riemannian submersion is harmonic,

$$\varphi_r = \pi \circ \psi_r \quad \mathbb{C}P^1 \to \mathbb{H}P^n$$

will be a minimal surface for all r. To understand why this is true using our methods, one must introduce the alternative twistor bundle

$$\widetilde{T} = \frac{Sp(n+1)}{U(n) \times Sp(1)}$$

with fibre $Sp(n)/U(n)$ over $\mathbb{H}P^n$ (see the table) discussed by Glazebrook [G]. Applying the same techniques as above, it can be shown that there exist $\widetilde{\psi}_r : \mathbb{C}P^1 \to \widetilde{T}$ with $\pi \circ \widetilde{\psi}_r = \varphi_r$ and this time

$\widetilde{\psi}_1$	horizontal holomorphic
$\widetilde{\psi}_3,\dots,\widetilde{\psi}_{2n+1}$	J_2-holomorphic, not horizontal

Consequently φ_r must be harmonic (by theorem 3.5) and conformal, i.e. minimal, for all r. Every φ_r is also real isotropic, although this is less obvious.

Other quaternionic Kähler manifolds include Grassmannians $\widetilde{G}_4(\mathbb{R}^{m+4})$ and $G_2(\mathbb{C}^{n+1})$. The twistor bundle of the latter with fibre $\mathbb{C}P^1$ can be identified with the fibration π'' in (6.2). Consequently the transform φ'' of a conformal harmonic map φ $M \to \mathbb{C}P^n$ is necessarily inclusive. In terms of subspaces of $\mathbb{C}^{n+1}$ we have $\varphi'' = \varphi \oplus \varphi'$. It follows from the remarks following corollary 6.2 that any inclusive harmonic map $f : S^2 \to G_2(\mathbb{C}^{n+1})$ has the form

$$f = h^{(r)} \oplus h^{(r+1)},$$

for some holomorphic map $h : S^2 \to \mathbb{C}P^n$, and some integer r, $0 \leqslant r \leqslant n$ (if $r = n$, $r+1$ must be replaced by 0). This is the basis of a classification of all harmonic maps $S^2 \to G_2(\mathbb{C}^4)$ given by Ramanathan [R], whose work has been extended by Chern & Wolfson to cover $G_2(\mathbb{C}^{n+1})$ for all n.

In another direction, Burstall [B1] has shown that a harmonic map from S^2 into any complex Grassmannian is the projection of a J_2-holomorphic curve in a suitable flag manifold. Such curves can be built up from complex analytic data by generalizing the construction underlying corollary 6.2.

7. BUNDLES OF PARTIALLY COMPLEX STRUCTURES

In this section we wish to drop the assumption that the targent manifold be even-dimensional. To do this one can consider complex structures only on subspaces of the tangent space. Let W be a real vector space of dimension $k = 2n+m$. A partially complex structure on W consists of a subspace W_1 of W of dimension $2n$, and an endomorphism J of W_1 with $J^2 = -1$. In the presence of an inner product on W satisfying

$$h(JX,JY) = h(X,Y), \quad X,Y \in W_1,$$

J can be extended to a transformation of W by setting it equal to zero on the orthogonal complement $W_1^{\perp}$ of W_1. Then

$$J^3 + J = 0, \tag{7.1}$$

and an endomorphism satisfying (7.1) is called an f-structure [Y;Ra].

Now let N be an oriented Riemannian manifold of dimension $k = 2n+m$. Following Rawnsley, one consider the associated bundle

$$F_n(N) = N \times_{SO(k)} \frac{SO(2n+m)}{U(n) \times SO(m)},$$

whose fibre parametrizes f-structures on T_xN for which $\dim W_1 = 2n$. Note that for $n \geqslant 2$ and $m \geqslant 1$, this fibre is precisely the twistor bundle of the real Grassmannian $\tilde{G}_m(\mathbb{R}^k)$, whereas $F_n(N) = S$ when $m = 0$. It is shown in [Ra] that many results concerning S are also applicable to $F_n(N)$. We shall focus on the other extreme case, namely $n = 1$.

The twistor space $F_1(N)$ can be identified with the bundle $\tilde{G}_2(TN)$ whose fibre at $x \in N$ is the Grassmannian $\tilde{G}_2(T_xN)$ of oriented real 2-dimensional subspaces of T_xN. Just as for S in (3.1), there is a splitting

$$T(\tilde{G}_2(TN)) = H \oplus V;$$

this time H does not carry an almost complex structure, but it does contain a tautologous subbundle H_1 whose fibre at a point of $\tilde{G}_2(TN)$ is the corresponding oriented 2-plane. Combining the natural almost complex structure on H_1 with the one on V induced from (5.2) gives two distinct endomorphisms J_1, J_2 of the subbundle $\Pi = H_1 \oplus V$ with $J_a^2 = -1$. They can be extended to f-structures by setting them equal to zero on the orthogonal complement of H_1 in H. Exactly one of these f-structures, by definition J_1, is integrable, that is to say the i-eigenbundle of J_1 is closed under Lie bracket and makes $\tilde{G}_2(TN)$ what is known as a <u>CR manifold</u>. The fact that J_1 is always integrable in this context follows because the relevant horizontal space has only one complex dimension, so there is no curvature obstruction.

The advantage of using $\tilde{G}_2(TN)$ is that any conformal immersion $\varphi : M \to N$ has a "Gauss lift"

$$\tilde{\varphi} : M \to \tilde{G}_2(TN)$$

for which $\tilde{\varphi}(x)$ is the oriented subspace $\varphi_*(T_x M)$. There is an obvious way of defining holomorphic curves in $\tilde{G}_2(TN)$; the map $\tilde{\varphi}$ is said to be J_a-<u>holomorphic</u> if

$$\tilde{\varphi}_*(TM) \subset \Pi \quad \text{and} \quad \tilde{\varphi}_* \circ J^M = J_a \circ \tilde{\varphi}_*.$$

<u>Theorem 7.1</u> A conformal map $\varphi : M \to N^k$, $k \geqslant 3$, is totally umbilic iff $\tilde{\varphi}$ is J_1-holomorphic, and harmonic iff $\tilde{\varphi}$ is J_2-holomorphic.

<u>Proof</u> First we remark that if φ is either totally umbilic or harmonic then $\tilde{\varphi}$ is well-defined even at points where $\varphi_* = 0$ (see proposition 2.2). The submanifold $\tilde{\varphi}(M)$ plays the same role as $i(S'(\varphi))$ in section 4, so it suffices to modify the proofs of theorem 4.1 and corollary 4.3.

The lift $\tilde{\varphi}$ determines an orthogonal sum

$$\varphi^{-1}TN = T_1 \oplus T_1^{\perp}$$

where $T_1 \cong \tilde{\varphi}^{-1}H_1$ is spanned generically by the real and imaginary components of $\delta\varphi$. The second fundamental form of T_1 in $\varphi^{-1}TN$ at a given $x \in M$ is an element of

$$\mathrm{Hom}(T_1, T_1^{\perp})_x = T_{\tilde{\varphi}(x)}\tilde{G}_2(T_xN).$$

In analogy with (4.2), $\tilde{\varphi}$ is J_1-holomorphic (respectively J_2-holomorphic iff $\delta^2\varphi$ (respectively $\overline{\delta}\delta\varphi$) is proportional to $\delta\varphi$. ■

Given a conformal immersion $\varphi: M \to N$, one can also interpret $\tilde{\varphi}$ as a section of the pullback $\varphi^{-1}\tilde{G}_2(TN)$. Theorem 7.1 is just a fancy way of describing when this "Gauss section" is holomorphic with respect to natural complex analyic structures on $(\varphi^{-1}TN)^{\mathbb{C}}$ (cf. (4.2)). For related properties of Gauss sections, see [E].

When $N = \mathbb{R}^k$ is Euclidean space, $\tilde{G}_2(TN)$ is simply a product $N \times \tilde{G}_2(\mathbb{R}^k)$, and if π_2 denotes projection onto the second factor, $\gamma_\varphi = \pi_2 \circ \tilde{\varphi}$ is the ordinary Gauss map of φ. Since γ_φ is holomorphic (respectively antiholomorphic) iff $\tilde{\varphi}$ is J_1-holomorphic (respectively J_2-holomorphic), one obtains the following well-known result [Ch_1;HO]:

<u>Corollary 7.2</u> A conformal immersion $\varphi : M \to \mathbb{R}^k$ is totally umbilic iff γ_φ is holomorphic, and harmonic iff γ_φ is antiholomorphic.

The isomorphism (5.2) gives

$$\tilde{G}_2(TN) \cong \{[X] \in P((TN)^{\mathbb{C}}) : h(X,X) = 0\},$$

realizing each fibre as a quadric of null vectors. This description depends only on the conformal class of the metric h of N, and likewise the CR structure defined by J_1 is conformally invariant. When $\dim N = 3$, the CR manifold $(\tilde{G}_2(TN), J_1)$ has nondegenerate Levi form with eigenvalues of opposite signs, and LeBrun [Le] proves that it can be realized as a real hypersurface of a complex 3-manifold iff N is conformally equivalent to a real analytic Riemannian manifold. The case $\dim N > 3$ has been considered by Rossi [Ro].

When N is 3-dimensional, $\tilde{G}_2(TN)$ may also be identified with the sphere bundle $S(TN)$ of unit tangent vectors, so that $\tilde{\varphi}$ becomes the normal vector field to $\varphi(M)$. A result of Stellmacher [ST] asserts that any Riemannian 3-manifold with totally umbilic surfaces through each point normal to each direction is conformally flat. This is then consistent with results of Bryant [Br_1] on the scarcity of holomorphic curves in a 5-dimensional Lorentzian CR manifold.

An exceptional example

The fact that any holomorphic curve in S^6 is automatically minimal (proposition 1.4, 2.1) make the above methods relevant to this example. The exceptional group G_2 is the group of algebra automorphisms of the Cayley numbers $\mathbb{O}$, and acts by orthogonal transformations on $\mathrm{Im}\,\mathbb{O} = \mathbb{R}^7$. This is the basic faithful representation of G_2, and since the isotropy subgroup of a vector $X \in \mathbb{R}^7$ is isomorphic to $SU(3)$, one obtains $S^6 \cong G_2/SU(3)$.

If $X \in \mathbb{R}^7$ is a unit vector, the orthogonal complement $(\mathbb{R}X)^\perp$ becomes an $SU(3)$-module, so X determines an f-structure on $\mathbb{R}^7$. This is in accordance with the inclusion

$$G_2/_{SU(3)} \hookrightarrow SO(7)/_{U(3)} .$$

Now consider $\mathbb{R}^7$ as a manifold with its standard Riemannian metric h, and flat Levi-Civita connection ∇. We have shown that the unit tangent bundle $\mathcal{U} = \mathbb{R}^7 \times S^6$ is naturally a subbundle of $F_3(\mathbb{R}^7)$. This fact can be used to carry out the following construction of Calabi [Ca_1].

Let P be any real hypersurface of $\mathbb{R}^7$, and let ∇^P denote the induced connection. The Gauss map $\gamma : P \to S^6$ is the composition $\pi_2 \circ \upsilon$, where $\upsilon : P \to \mathcal{U}$ is the normal vector field and π_2 is the projection. By above, the lift υ defines an almost Hermitian structure on P and so a fundamental 2-vector F and a decomposition

$$(TP)^{\mathbb{C}} = T^{1,0} \oplus T^{0,1}.$$

Using the SU(3)-isomorphism $T^{2,0} \cong T^{0,1}$, the components (1.6) of the tensor $\nabla^P F$ are

$$D_1 F \in T^{1,1} \oplus T^{1,1}$$

(7.2)

$$D_2 F \in (T^{0,1} \otimes T^{0,1}) \oplus (T^{1,0} \otimes T^{1,0}).$$

We leave the reader to verify that $\nabla^P F$ can also identified with the second fundamental form $-\nabla\upsilon$ of P in $\mathbb{R}^7$, which belongs to

(7.3) $$S^2(T^*P)^{\mathbb{C}} = T^{1,1} \oplus (S^2T^{1,0} \oplus S^2T^{0,1}).$$

Comparing (7.2) with (7.3), $\mathrm{tr}(D_2F) = 0$, so P is always cosymplectic [Gr_2, theorem 6.8]. Moreover $\gamma : M \to S^6$ is a mapping between two almost complex manifolds, and the same reasoning as in proposition 3.2 gives

<u>Proposition 7.3</u> The G_2-induced almost Hermitian structure on P satisfies $D_2F = 0$ iff γ is holomorphic, and $D_1F = 0$ iff γ is anti-holomorphic.

If $D_2F = 0$ so that P is (1,2)-symplectic, it is known that P is locally isometric to one of $\mathbb{R}^6$, S^6, $S^2 \times \mathbb{R}^4$ [YS, Gr_2]. In the third case, the almost Hermitian manifold is really the twistor space S of $\mathbb{R}^4$ with its J_2 structure. On the other hand if $D_1F = 0$, so that P is minimal, the fact that P is a complex manifold and S^6 is not implies that γ cannot be a local diffeomorphism anywhere. Moreover by [Gr_2], S^6 admits no holomorphic submanifolds of real dimension 4, so $\gamma(P)$ is a holomorphic curve in S^6. A class of examples are of the form $P = \mathbb{R}^4 \times M$, where $\mathbb{R}^3$ is a subspace of $\mathbb{R}^7$ such that $\gamma(P) = \mathbb{R}^3 \cap S^6 = S^2$ is holomorphic and $M \subset \mathbb{R}^3$ is minimal $[Ca_2]$; the complex structure induced on P is nontrivial. Finally note that in both the cases of proposition 7.3, γ is harmonic and P has constant mean curvature, thereby illustating the theorem of Ruh & Vilms [RV].

A study of holomorphic curves in S^6 has been made by Bryant $[Br_2]$; we explain his ideas briefly. Let $\varphi : M \to S^6$ be a holomorphic map from a Riemann surface. Because S^6 is not a complex manifold, φ will not necessarily be real isotropic. However it is so to the first order, indeed from (1.4), (1.5), (1.8), we have $(\delta^2\varphi)^{0,1} = \eta_{\delta\varphi}(\delta\varphi) = 0$, and

$$\delta^2\varphi \in \Gamma(M, \varphi^{-1}T^{1,0}S^6).$$

Giving the vector bundle $\varphi^{-1}T^{1,0}S^6$ over M its natural complex analytic structure, one checks that the forms

$$\begin{aligned}
\lambda_1 &= dz \otimes \delta\varphi \\
\lambda_2 &= dz^3 \otimes (\delta\varphi \wedge \delta^2\varphi) \\
\lambda_3 &= dz^6 \otimes (\delta\varphi \wedge \delta^2\varphi \wedge (\delta^3\varphi)^{1,0})
\end{aligned}$$

are all well-defined and complex analytic. Now λ_3 is identically zero iff φ is real isotropic (φ is then said to have <u>null torsion</u> in $[Br_2]$). For $(\delta^3\varphi)^{0,1} = \eta_{\delta\varphi}\delta^2\varphi$ is proportional to

$$\delta\varphi \wedge \delta^2\varphi \in T^{2,0} \cong T^{0,1},$$

and $h(\delta^3\varphi, \delta^3\varphi) = 0$ iff $(\delta^3\varphi)^{1,0}$ is a linear combination of $\delta\varphi$, $\delta^2\varphi$.

If φ is real isotropic, then by theorem 4.5 the almost complex structure for which $\delta\varphi$, $\delta^2\varphi$, $\delta^3\varphi$ are (1,0)-vectors defines a horizontal holomorphic curve ψ in S. However this almost complex structure is precisely J_W as defined in (6.1) where W is the real subspace generated by $\delta\varphi$, $\delta^2\varphi$. Thus ψ lies in the twistor space

$$G_2(T^{1,0}S^6) \cong G_2/U(2) \cong Q^5$$

which is a J_1-holomorphic submanifold of $\mathcal{S}(S^6)$ [OR]. Bryant calls ψ the binormal mapping, and uses it to prove that there exists a holomorphic map $M \to S^6$ for any compact Riemann surface M.

In the above example, the lift ψ also coincides with Chern's directrix ζ_φ [Ch_2], which can be defined for any full real isotropic harmonic map φ $M \to S^{2n}$ as follows. The component of $\delta^n\varphi$ orthogonal to $\delta\varphi, \delta^2\varphi, \ldots, \delta^{n-1}\varphi$ generates a J_1-holomorphic lift $M \to \tilde{G}_2(TS^{2n})$, and ζ_φ is the composition of this with the natural projection

$$\tilde{G}_2(TS^{2n}) \cong \frac{SO(2n+1)}{SO(2) \times SO(2n-2)} \to \frac{SO(2n+1)}{SO(2) \times SO(2n-1)} \cong Q^{2n-1} \subset \mathbb{C}P^{2n}.$$

The correspondence $\varphi \leftrightarrow \zeta_\varphi$ is 2 . 1, just as in theorem 4.5 [B].

8. FOUR DIMENSIONS

From now on we suppose that N is an oriented 4-dimensional Riemannian manifold, and refer the reader to [AHS; S_2] for relevant background material. Any point of $\tilde{G}_2(TN)$, i.e. an oriented real 2-dimensional subspace $W \subset T_xN$, gives rise to two almost complex structures J_+, J_- as follows. Let $\{e_1, e_2, e_3, e_4\}$ be an oriented orthonormal basis of T_xN with $W = \text{span}\{e_1, e_2\}$ and set

$$J_\pm e_1 = e_2 . \quad J_\pm e_3 = \pm e_4$$

thus $J_+ \geqslant 0$, whereas $J_- \leqslant 0$. This observation leads one to consider not only $S = S_+$, but also the bundle S_- of Riemannian almost complex structures with reversed orientation. The distinction between S_+ and S_- is best appreciated in terms of 2-vectors. The SO(4)-structure on TN defines the Hodge star operator

$$* : \Lambda^2 TN \to \Lambda^2 TN$$

with $*^2 = 1$, and there is a spectral decomposition

$$\Lambda^2 TN = \Lambda^2_+ TN \oplus \Lambda^2_- TN \tag{8.1}$$

corresponding to the eigenvalues $+1, -1$. Let $S(\Lambda^2_\pm TN)$ denote the associated bundles of unit vectors.

<u>Proposition 8.1</u> $S_\pm \cong S(\Lambda^2_\pm TN)$.

<u>Proof</u> Given $J \in (S_\pm)_x$, there exists an oriented orthonormal basis with $Je_1 = e_2$ and $Je_3 = \pm e_4$. The fundamental 2-vector $F = e_1 \wedge e_2 \pm e_3 \wedge e_4$ satisfies $*F = \pm F$, and has norm 1 with an appropriate convention for defining the induced metric on $\Lambda^2 TN$.

Conversely any $F \in S(\Lambda^2_{\pm}TN)$ equals $e_1 \wedge e_2 \pm e_3 \wedge e_4$ for some basis and defines an element of $S_{\pm}$. This is because the action of SO(4) on (8.1) induces a double covering

$$SO(4) \to SO(3) \times SO(3)$$

which is represented by mapping an oriented orthonormal basis $\{e_1, e_2, e_3, e_4\}$ of TN to the oriented orthonormal bases

$$\{e_1 \wedge e_2 \pm e_3 \wedge e_4,\ e_1 \wedge e_3 \pm e_4 \wedge e_2,\ e_1 \wedge e_4 \pm e_2 \wedge e_3\}$$

of $\Lambda^2_{\pm}TN$. ∎

The above arguments show that the fibre $\tilde{G}_2(\mathbb{R}^4) \cong Q^2$ of $\tilde{G}_2(TN)$ is isomorphic to $S^2 \times S^2$, and that there is the following diagram in which every diagonal fibre is S^2.

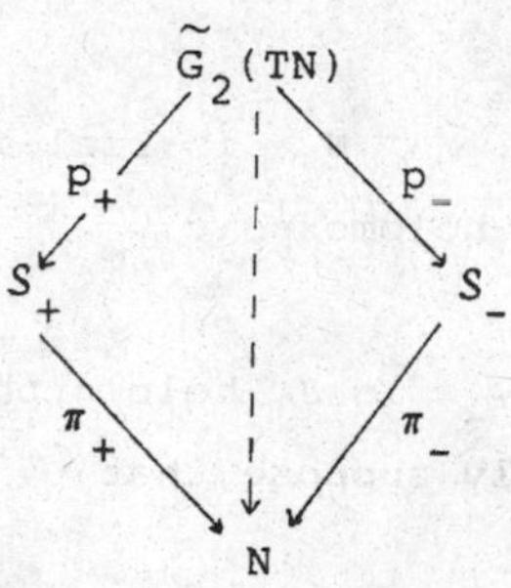

The Gauss lift of a conformal immersion $\varphi : M \to N$ now has the form

$$(8.2) \qquad \tilde{\varphi} = (\tilde{\varphi}_+, \tilde{\varphi}_-),$$

where $\tilde{\varphi}_{\pm} = p_{\pm} \circ \tilde{\varphi}$. If ω triviales $\Lambda^2 T_x M$, then $\tilde{\varphi}_{\pm}(x)$ is the normalization of the projection of $\varphi_* \omega$ in $\Lambda^2_{\pm}TN$. Alternatively $\tilde{\varphi}_{\pm}(x)$ is the unique almost complex structure in $S_{\pm}$ making $\delta\varphi(x)$ a (1,0)-vector, and is essentially $S'(\varphi)$ in the language of theorem 4.1.

Of course proposition 3.1 is equally valid for S_+ and S_-, and a conformal harmonic map $\varphi : M \to N$ will be harmonic iff $\tilde{\varphi}$ is J_2-holomorphic, or iff $\tilde{\varphi}_+$ is J_2-holomorphic, or iff $\tilde{\varphi}_-$ is J_2-holomorphic. Accordingly we shall now forget $\tilde{G}_2(TN)$, and concentrate on the almost complex manifolds S_+, S_-. Given a smooth J_2-holomorphic curve $\psi : M \to S_\pm$, $\varphi = \pi_\pm \circ \psi$ is minimal. Provided that ψ is non-vertical (i.e. not contained in a fibre), φ_* has only isolated zeros and $\psi = \tilde{\varphi}_\pm$ is the unique J_2-holomorphic curve projecting to φ; thus

<u>Theorem 8.2</u> There is a 1 . 1 . 1 correspondence between minimal surfaces in N^4, nonvertical J_2-holomorphic curves in S_+, and non-vertical J_2-holomorphic curves in S_-.

Moving on to J_1, by theorem 7.1 and (8.2) a conformal map $\varphi : M \to N^4$ is totally umbilic iff both $\tilde{\varphi}_+$ and $\tilde{\varphi}_-$ are J_1-holomorphic On the other hand

<u>Proposition 8.3</u> A map $\varphi : M \to N^4$ is real isotropic iff locally at least one of $\tilde{\varphi}_+, \tilde{\varphi}_-$ is J_1-holomorphic.

<u>Proof</u> If either $\tilde{\varphi}_+$ or $\tilde{\varphi}_-$ is J_1-holomorphic, φ is isotropic by proposition 4.4. Conversely suppose that

$$h(\delta\varphi, \delta\varphi) = 0 = h(\delta^2\varphi, \delta^2\varphi) \tag{8.3}$$

everywhere so that in addition $h(\delta\varphi, \delta^2\varphi) = 0$; for $\dim N = 4$, (4.3) and (8.3) are actually equivalent. At each point, unless $\delta^2\varphi$ is proportional to $\delta\varphi$, $\{\delta\varphi, \delta^2\varphi\}$ spans the (1,0)-space of some almost complex structure J. Now $J \in S_+$ or $J \in S_-$, and from the proof of corollary 4.3, $\tilde{\varphi}_+$ or $\tilde{\varphi}_-$ respectively is J_1-holomorphic. This argument shows that each point of M lies in the closure of a neighbourhood on which at least one of $\tilde{\varphi}_+, \tilde{\varphi}_-$ is J_1-holomorphic; this is the meaning of the proposition . ■

If a map $\psi : M \to S_+$ is both J_1 and J_2-holomorphic, then from the definitions it is tangent to the horizontal distribution H in (3.1). The existence of such a horizontal map imposes stringent conditions on the curvature of the bundle S_+ and so on the Riemannian curvature tensor of N. In the other direction, if N is an anti-self-dual Einstein 4-manifold then the distribution $(T^{1,0})^h$ of horizontal (1,0) vectors in S_+ is complex analytic [S_2, theorem 10.1], and there is no local obstruction to the existence of horizontal holomorphic curves in S_+. With the same hypotheses on N (or with anti-self-dual replaced by self-dual) proposition 8.3 holds globally, and therefore at least one of the Gauss lifts $\tilde{\varphi}_+, \tilde{\varphi}_-$ of a real isotropic harmonic map $\varphi : M \to N$ must be everywhere horizontal. Both of them are horizontal iff φ is totally geodesic.

Concluding examples

1. The simplest 4-manifold for which the total spaces S_+ and S_- are different is the complex projective plane $\mathbb{C}P^2$. Adopting the standard orientation, $\mathbb{C}P^2$ is self-dual and

$$S_- \cong P(T^{1,0}\mathbb{C}P^2) \cong \frac{U(3)}{U(1) \times U(1) \times U(1)} .$$

Moreover $S_+ \cong P(\kappa \oplus \underline{\mathbb{C}})$ where $\kappa = \Lambda^2(T^{1,0}\mathbb{C}P^2)^*$ is the canonical bundle and $\underline{\mathbb{C}}$ the trivial bundle. Thus S_+ is not a homogeneous space, but has two natural sections corresponding to the complex structure of $\mathbb{C}P^2$ and its conjugate. More details of these and subsequent assertions are contained in [ES_2].

Let $\varphi : M \to \mathbb{C}P^2$ be a conformal harmonic map, i.e. a branched minimal immersion. Then $\tilde{\varphi}_+$ is horizontal iff φ is totally real, holomorphic or antiholomorphic, whereas $\tilde{\varphi}_-$ is horizontal iff φ is the transform of some holomorphic map (see corollary 6.2). The last three cases occur iff φ is complex isotropic (6.4), and are essentially classified. Furthermore if φ is complex isotropic, but neither holomorphic nor antiholomorphic then it cannot be an embedding [ETG, proposition 3.5]. Now any harmonic map $S^2 \to \mathbb{C}P^n$ is necessarily complex isotropic, as is any harmonic map of nonzero degree from the torus T^2 into $\mathbb{C}P^n$ [EW, section 7]. There do exist totally real minimal immersions $T^2 \to \mathbb{C}P^2$ [L_1, N], note that although these are not complex isotropic, they are real isotropic. In all cases $\tilde{\varphi}_-$ is itself a minimal surface in the flag manifold S_- by theorem 6.3.

2. Micallef [M] has shown that a large class of complete oriented <u>stable</u> minimal surfaces $\varphi : M \to \mathbb{R}^4$, including those of finite total curvature, are holomorphic with respect to some orthogonal complex structure on $\mathbb{R}^4$. This is equivalent to saying that $\tilde{\varphi}_+$ or $\tilde{\varphi}_-$ is horizontal, and that φ is real isotropic. A corresponding theorem is given when M has genus zero in $\mathbb{R}^k$ for any k.

3. Bryant [Br_3] calls real isotropic harmonic maps into the sphere S^4 <u>superminimal</u>; the prefix "super" refers to the vanishing of the holomorphic quartic differential $h(\delta^2\varphi, \delta^2\varphi)dz^4$. Now S^4 is conformally flat, and its twistor bundles S_+, S_- are isomorphic as homogeneous spaces: SO(5) acts transitively on each, and the two isotropy subgroups $U(2)_+, U(2)_-$ are conjugate in SO(5). Moreover $S^4 \cong \mathbb{H}P^1$ and

$$S_\pm \cong \frac{SO(5)}{U(2)} \cong \frac{Sp(2)}{U(1) \times Sp(1)} \cong \mathbb{C}P^3$$

(cf. (6.5)). The isotropy representation has exactly two irreducible real components, so there are $2^2 = 4$ SO(5)-invariant almost complex structures on $\mathbb{C}P^3$ [BH]; these are $\pm J_1, \pm J_2$, J_1 being the standard complex structure. In [Br_3] a horizontal holomorphic embedding $M \to \mathbb{C}P^3$ is constructed for any compact Riemann surface M; this then projects to a superminimal immersion $M \to S^4$.

Both Gauss lifts $\tilde{\varphi}_+, \tilde{\varphi}_-$ of a minimal surface $\varphi : M \to S^4$ are themselves minimal surfaces in $\mathbb{C}P^3$ relative to the Fubuni-Study metric (theorem 6.3), in analogy to the theorem of Ruh & Vilms [RV]. Using the Riemannian submersion equations, one can show that both

Full minimal surfaces in $\mathbb{C}P^3 = S_+$

Shading = horizontal holomorphic ones

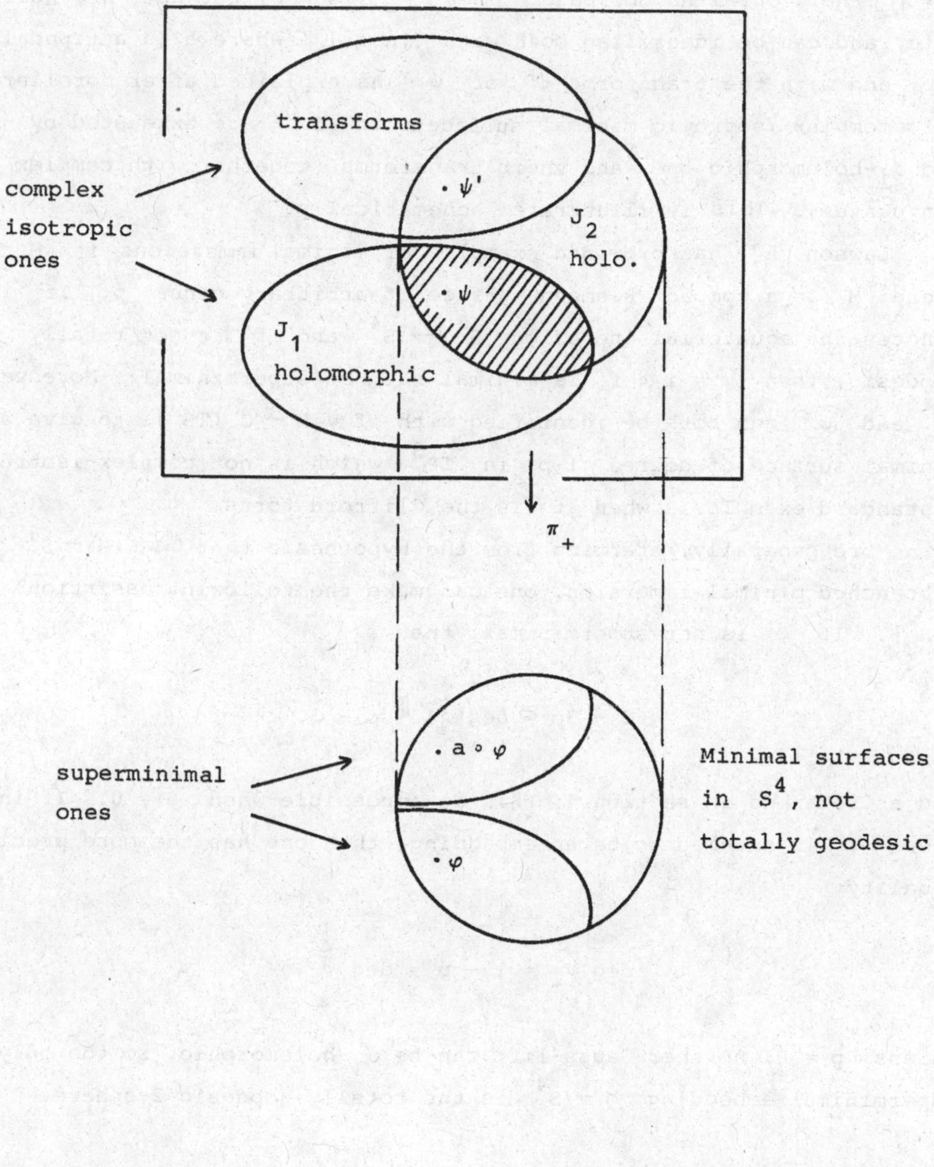

$\tilde{\varphi}_+$ and $\tilde{\varphi}_-$ are complex isotropic iff φ is superminimal.

Let $\varphi\colon M \to S^4$ be superminimal, not totally geodesic, with $\psi = \tilde{\varphi}_+$ horizontal holomorphic. Then $\tilde{\varphi}_-$ is J_2, but not J_1, holomorphic, and can be identified both with $(\widetilde{a \circ \varphi})_+$ where a is antipodal map, and with the transform ψ' of ψ. As explained after corollary 6.2, complex isotropic minimal surfaces in $\mathbb{C}P^3$ are exhausted by the J_1-holomorphic ones and their transforms (together with complex conjugates). This is illustrated schematically.

Lawson [L_2] has provided examples of minimal immersions $f : M \to S^3$, where M is a compact Riemann surface of arbitrary genus p. If ι denotes the equatorial inclusion $S^3 \hookrightarrow S^4$ and f is not totally geodesic, then $\varphi = \iota \circ f$ is minimal but not superminimal. Moreover $\tilde{\varphi}_+$ and $\tilde{\varphi}_-$ can both be identified with $\tilde{f}\colon M \to \tilde{G}_2(TS^3)$ to give a minimal surface of degree $1-p$ in $\mathbb{C}P^3$ which is not complex isotropic. A standard example is when f is the Clifford torus.

More generally, starting from the hypothesis that $\varphi\colon M \to S^4$ is a branched minimal immersion, one can make the following assertions [ES_2]. If φ is not superminimal, then

$$3 - 3p \leqslant \deg(\tilde{\varphi}_\pm) \leqslant p - 1,$$

and as remarked in section 4, this is impossible when $p = 0$. If instead φ is assumed to be an embedding, then one has the more precise equality

$$\deg \tilde{\varphi}_+ = 1 - p = \deg \tilde{\varphi}_-.$$

Unless $p = 0$, neither Gauss lift can be J_1-holomorphic, so the only superminimal embedding $M \to S^4$ is the totally geodesic 2-sphere.

REFERENCES

[AHS] Atiyah, M.F., Hitchin, N.J., Singer, I.M.. Self-duality in four-dimensional Riemannian geometry, Proc. R. Soc. Lond Ser. A 362 (1978) 425-461.

[B] Barbosa, J.· On minimal immersions of S^2 into S^{2m}, Trans. Am. Math. Soc. 210 (1975) 75-106.

[BO] Bérard Bergery, L., Ochiai, T.· On some generalization of the construction of twistor space, in Global Riemannian Geometry, edited by T.J. Willmore and N.J. Hitchin, Ellis Horwood, 1984.

[BH] Borel, A., Hirzebruch, F.: Characteristic classes and homogeneous spaces I, Am. J. Math. 80 (1958) 458-538.

[Br_1] Bryant, R.L.: Holomorphic curves in Lorentzian CR manifolds, Trans. Am. Math. Soc. 272 (1982) 203-221.

[Br_2] Bryant, R.L.. Special structures and the octonians, J Diff. Geom 17 (1982) 185-232.

[Br_3] Bryant, R.L.. Conformal and minimal immersions of compact surfaces into the 4-sphere, J. Diff. Geom. 17 (1982) 455-473.

[Br_4] Bryant, R.L.. Lie groups and twistor spaces, preprint.

[Bu] Burns, D.. Harmonic maps from $\mathbb{C}P^1$ to $\mathbb{C}P^n$, in Harmonic Maps, Lecture Notes in Mathematics 949, Springer, 1981.

[Bl] Burstall, F.· Twistor fibrations of flag manifolds over Grassmannians.

[Ca_1] Calabi, E.. Construction and properties of some 6-dimensional almost complex manifolds, Trans. Am. Math. Soc. 87 (1958) 407-438.

[Ca_2] Calabi, E.: Minimal immersions of surfaces in Euclidean spheres J. Diff. Geom. 1 (1967) 111-125.

[Ca_3] Calabi, E.: Quelques applications de l'analyse complexe aux surfaces d'aire minime, in Topics in Complex manifolds, University of Montreal, 1967.

[C] Chen, B.-Y.: Geometry of Submanifolds, Marcel Dekker, 1973.

[Ch_1] Chern, S.S.: Minimal surfaces in Euclidean space of N dimensions, in Differential and Combinatorial Topology, edited by S.S. Cairns, Princeton University Press, 1965.

[Ch_2] Chern S.S.: On the minimal immersions of the two-sphere in a space of constant curvature, in Problems in Analysis, edited by R.C. Gunning, Princeton University Press, 1970.

[DZ] Din, A.M., Zakrzewski, W.J.: General classical solutions in the $\mathbb{C}P^{n-1}$ model, Nuclear Phys. B 174 (1980) 397-406.

[E] Eells, J.: Gauss maps of surfaces, in Perspectives in Mathematics, Anniversary of Oberwolfach, Birkhäuser, 1984.

[EL] Eells, J., Lemarie, L.: A report on harmonic maps, Bull. Lond. Math. Soc. 10 (1978) 1-68.

[ES_1] Eells, J., Salamon, S.: Constructions twistorielles des applications harmoniques, C. R. Acad. Sc. Paris 296 (1983) 685-687.

[ES_2] Eells, J., Salamon, S.: Twistorial construction of harmonic maps of surfaces into four-manifolds, Ann.Sc.Norm.Sup. Pisa.

[ESa] Eells, J., Sampson, J.H.. Harmonic mappings of Riemannian manifolds, Amer. J. Math. 86 (1964) 109-160.

[EW] Eells, J., Wood, J.C.. Harmonic maps from surfaces to complex projective spaces, Adv. in Math. 49 (1983) 217-263.

[ErW] Erdem, S., Wood; J.C.: On the construction of harmonic maps into a Grassmannian, J. Lond. Math. Soc. 28 (1983) 161-174.

[ETG] Eschenburg, J.H., Tribuzy, R.de A., Guadalupe, I.V.: The fundamental equations of minimal surfaces in $\mathbb{C}P^2$, preprint.

[FI] Fukami, T., Ishihára, S.: Almost Hermitian structure on S^6, Tôhoku Math. J. 7 (1955) 151-156.

[GS] Glaser, V., Stora, R.: Regular solutions of the $\mathbb{C}P^n$ models and further generalizations, preprint.

[G] Glazebrook, J.: The construction of a class of harmonic maps to quaternonic projective space, to appear in J. Lond. Math. Soc.

[Gr_1] Gray, A.: Minimal varieties and almost hermitian submanifolds, Michigan Math. J. 12 (1965) 273-287.

[Gr_2] Gray, A.: Vector cross products on manifolds, Trans. Am. Math. Soc. 141 (1969) 465-504.

[Gr_3] Gray, A.: Almost complex submanifolds of the six sphere, Proc. Am. Math. Soc. 20 (1969) 277-279.

[Gr_4] Gray, A.. Riemannian manifolds with geodesic symmetries of order 3, J. Diff. Geom. 7 (1972) 343-369.

[GH] Gray, A., Hervella, L.M.. The sixteen classes of almost Hermitian manifolds and their linear invariants, Ann. Mat. Pura Appl. 123 (1980) 35-58.

[GOR] Gulliver, R.D., Osserman, R., Royden, H.L.: A theory of branched immersions of surfaces, Am. J. Math. 95 (1973) 750-812.

[HO] Hoffman, D.A., Osserman, R.: The geometry of the generalized Gauss map, Mem. Am. Math. Soc. 236 (1980).

[JR] Jensen, G.R., Rigoli, M.: Minimal surfaces in spheres by the the method of moving frames, preprint.

[KM] Koszul, J.-L., Malgrange, B.. Sur certaines structures fibrées complexes, Arch. Math. 9 (1958) 102-109.

[K] Kotō, S.: Some theorems on almost Kählerian spaces, J. Math. Soc. Japan 12 (1960) 422-433.

[L_1] Lawson, H.B.: Rigidity theorems in rank-1 symmetric spaces, J. Diff. Geom. 4 (1970) 349-357.

[L_2] Lawson, H.B.: Complete minimal surfaces in S^3, Ann. Math. 92 (1970) 335-374.

[L_3] Lawson, H.B.: Lectures on Minimal Submanifolds, 2nd edition Publish or Perish, 1980.

[Le] LeBrun, C.R.. Twistor CR manifolds and 3-dimensional conformal geometry, Trans. Am. Math. Soc. 284 (1984) 601-616.

[Li] Lichnerowicz, A.: Applications harmoniques et variétés kählériennes, Symposia Mathematica 3 (1980) 341-402.

[M] Micallef, M.J.: Stable minimal surfaces in Euclidean space, J. Diff. Geom. 19 (1984) 57-84.

[N] Naitoh, H.. Isotropic submanifolds with parallel second fundamental form in $P^m(c)$, Osaka J. Math. 18 (1981) 427-464.

[NW] Nijenhuis, A., Woolf, W.B.: Some integration problems in almost-complex and complex manifolds, Ann. Math. 77 (1963) 424-489.

[OR] O'Brian, N.R., Rawnsley, J.H.: Twistor spaces, to appear in Ann. Global Analysis and Geometry.

[O] O'Neill, B.: The fundamental equations of a submersion, Mich. Math. J. 10 (1963) 335-339.

[R] Ramanathan, J.. Harmonic maps from S^2 to $G_{2,4}$, J. Diff. Geom. 19 (1984) 207-219.

[Ra] Rawnsley, J.H.. f-structures, f-twistor spaces and harmonic maps, in Geometry Seminar Luigi Bianchi, same volume.

[Ro] Rossi, H.: LeBrun's non-realizability theorem in higher dimensions, preprint.

[RV] Ruh, E.A., Vilms, J.. The tension field of the Gauss map, Trans. Amer. Math. Soc. 149 (1970) 569-573.

[S_1] Salamon, S.. Quaternionic Kähler manifolds, Invent. Math. 67 (1982) 143-171.

[S_2] Salamon, S.: Topics in four-dimensional Riemannian geometry, in Geometry Seminar Luigi Bianchi, Lecture Notes in Mathematics 1022, Springer, 1983.

[S_3] Salamon, S.. Quaternionic structures and twistor spaces, in Global Riemannian Geometry, edited by T.J. Willmore and N.J. Hitchin, Ellis Horwood, 1984.

[Sk] Skornyakov, I.A.. Generalized Atiyah-Ward bundles, Uspekhi Mat. Nauk. 37 (1982) 195-196.

[St] Stellmacher, K.L.. Geometrische Deutung konforminvarianter Eigen schaften des Riemannschen Raumes, Math. Ann. 123 (1959) 34-52.

[W] Wolf, J.A.: Complex homogeneous contact manifolds and quaternionic symmetric spaces, J. Math. Mech. 14 (1965) 1033-1047.

[WG] Wolf, J.A., Gray, A.: Homogeneous spaces defined by Lie group automorphisms I,II, J. Diff. Geom. 2 (1968) 77-159.

[Wo] Wolfson, J.G.. Minimal surfaces in complex manifolds, Ph.D. thesis, University of California, Berkeley, 1982.

[Wd] Wood, J.C.: Singularities of harmonic maps and applications of the Gauss-Bonnet formula, Am. J. Math. 99 (1977) 1329-1344

[Y] Yano, K.: On a structure defined by a tensor field f of type (1,1) satisfying $f^3 + f = 0$, Tensor 14 (1963) 99- 109.

[YS] Yano, K., Sumitomo, T.. Differential Geometry of hypersurfaces in a Cayley space, Proc. Roy. Soc. Edinburgh A 66 (1964) 216-231.

Vol 1008 Algebraic Geometry Proceedings, 1981 Edited by J Dolgachev V, 138 pages 1983

Vol 1009 T A Chapman, Controlled Simple Homotopy Theory and Applications III, 94 pages 1983

Vol 1010 J-E Dies, Chaînes de Markov sur les permutations IX, 226 pages 1983

Vol 1011 J M Sigal Scattering Theory for Many-Body Quantum Mechanical Systems IV, 132 pages 1983

Vol 1012· S Kantorovitz, Spectral Theory of Banach Space Operators. V, 179 pages. 1983

Vol 1013 Complex Analysis – Fifth Romanian-Finnish Seminar Part 1. Proceedings, 1981 Edited by C Andreian Cazacu, N Boboc, M Jurchescu and I Suciu XX, 393 pages 1983

Vol 1014 Complex Analysis – Fifth Romanian-Finnish Seminar Part 2 Proceedings, 1981 Edited by C Andreian Cazacu, N Boboc, M Jurchescu and I Suciu XX, 334 pages 1983.

Vol 1015 Equations différentielles et systèmes de Pfaff dans le champ complexe – II Seminar Edited by R Gérard et J P Ramis V, 411 pages 1983

Vol 1016 Algebraic Geometry Proceedings, 1982 Edited by M Raynaud and T Shioda VIII, 528 pages 1983.

Vol 1017 Equadiff 82 Proceedings, 1982 Edited by H W Knobloch and K Schmitt XXIII, 666 pages 1983

Vol 1018 Graph Theory, Łagów 1981 Proceedings, 1981 Edited by M Borowiecki, J W Kennedy and M M Sysło X, 289 pages 1983

Vol. 1019. Cabal Seminar 79–81 Proceedings, 1979–81 Edited by A S Kechris, D A Martin and Y N Moschovakis V, 284 pages 1983

Vol. 1020 Non Commutative Harmonic Analysis and Lie Groups Proceedings, 1982 Edited by J Carmona and M Vergne V, 187 pages 1983

Vol. 1021 Probability Theory and Mathematical Statistics Proceedings, 1982 Edited by K Itô and J V Prokhorov VIII, 747 pages 1983

Vol 1022: G Gentili, S Salamon and J-P Vigue Geometry Seminar "Luigi Bianchi", 1982 Edited by E. Vesentini VI, 177 pages 1983

Vol 1023 S McAdam, Asymptotic Prime Divisors IX, 118 pages 1983

Vol. 1024. Lie Group Representations I. Proceedings, 1982–1983 Edited by R Herb, R Lipsman and J Rosenberg. IX, 369 pages 1983

Vol 1025 D Tanré, Homotopie Rationnelle Modèles de Chen, Quillen, Sullivan X, 211 pages 1983

Vol 1026 W Plesken, Group Rings of Finite Groups Over p-adic Integers V, 151 pages 1983

Vol 1027 M Hasumi, Hardy Classes on Infinitely Connected Riemann Surfaces XII, 280 pages 1983

Vol 1028· Séminaire d'Analyse P Lelong – P Dolbeault – H Skoda Années 1981/1983 Edité par P Lelong, P Dolbeault et H Skoda VIII, 328 pages 1983

Vol. 1029 Séminaire d'Algèbre Paul Dubreil et Marie-Paule Malliavin Proceedings, 1982 Edité par M-P Malliavin V, 339 pages 1983

Vol. 1030 U Christian, Selberg's Zeta-, L-, and Eisensteinseries XII, 196 pages 1983

Vol 1031. Dynamics and Processes Proceedings, 1981 Edited by Ph. Blanchard and L Streit IX, 213 pages 1983

Vol 1032 Ordinary Differential Equations and Operators Proceedings, 1982 Edited by W N Everitt and R T Lewis XV, 521 pages 1983

Vol 1033 Measure Theory and its Applications Proceedings, 1982 Edited by J M Belley, J Dubois and P Morales XV, 317 pages 1983

Vol 1034 J Musielak, Orlicz Spaces and Modular Spaces V, 222 pages 1983

Vol 1035 The Mathematics and Physics of Disordered Media Proceedings, 1983 Edited by B D Hughes and B.W Ninham VII, 432 pages 1983

Vol 1036. Combinatorial Mathematics X Proceedings, 1982 Edited by L R A Casse XI, 419 pages 1983

Vol 1037 Non-linear Partial Differential Operators and Quantization Procedures Proceedings, 1981 Edited by S I Andersson and H-D Doebner VII, 334 pages 1983

Vol 1038. F Borceux, G Van den Bossche, Algebra in a Localic Topos with Applications to Ring Theory IX, 240 pages 1983

Vol 1039 Analytic Functions, Błażejewko 1982 Proceedings Edited by J Ławrynowicz X, 494 pages. 1983

Vol 1040 A Good, Local Analysis of Selberg's Trace Formula III, 128 pages 1983.

Vol 1041 Lie Group Representations II Proceedings 1982–1983 Edited by R Herb, S Kudla, R Lipsman and J Rosenberg. IX, 340 pages 1984

Vol 1042 A Gut, K D Schmidt, Amarts and Set Function Processes. III, 258 pages 1983

Vol 1043 Linear and Complex Analysis Problem Book Edited by V P Havin, S V Hruščev and N K Nikol'skii XVIII, 721 pages 1984

Vol 1044 E Gekeler, Discretization Methods for Stable Initial Value Problems VIII, 201 pages 1984

Vol 1045 Differential Geometry Proceedings, 1982 Edited by A M Naveira VIII, 194 pages 1984

Vol 1046 Algebraic K-Theory, Number Theory, Geometry and Analysis Proceedings, 1982. Edited by A Bak IX, 464 pages 1984

Vol 1047 Fluid Dynamics Seminar, 1982 Edited by H Beirão da Veiga VII, 193 pages 1984

Vol 1048. Kinetic Theories and the Boltzmann Equation Seminar, 1981 Edited by C Cercignani VII, 248 pages 1984.

Vol 1049 B Iochum, Cônes autopolaires et algèbres de Jordan VI, 247 pages 1984

Vol 1050 A Prestel, P Roquette, Formally p-adic Fields V, 167 pages 1984

Vol 1051 Algebraic Topology, Aarhus 1982 Proceedings Edited by I Madsen and B Oliver X, 665 pages 1984

Vol 1052 Number Theory Seminar, 1982 Edited by D V Chudnovsky, G.V Chudnovsky, H Cohn and M B Nathanson V, 309 pages 1984

Vol 1053 P Hilton, Nilpotente Gruppen und nilpotente Raume V, 221 pages 1984

Vol 1054 V Thomée. Galerkin Finite Element Methods for Parabolic Problems VII, 237 pages 1984

Vol 1055 Quantum Probability and Applications to the Quantum Theory of Irreversible Processes Proceedings, 1982 Edited by L Accardi, A Frigerio and V Gorini VI, 411 pages 1984

Vol 1056 Algebraic Geometry Bucharest 1982 Proceedings, 1982 Edited by L Bădescu and D Popescu. VII, 380 pages 1984

Vol 1057 Bifurcation Theory and Applications Seminar, 1983 Edited by L Salvadori VII, 233 pages 1984

Vol 1058 B Aulbach, Continuous and Discrete Dynamics near Manifolds of Equilibria. IX, 142 pages 1984

Vol 1059 Séminaire de Probabilités XVIII, 1982/83 Proceedings Edité par J Azéma et M Yor IV, 518 pages 1984

Vol 1060 Topology Proceedings, 1982 Edited by L D Faddeev and A A Mal'cev VI, 389 pages 1984

Vol 1061 Séminaire de Théorie du Potentiel Paris, No. 7 Proceedings Directeurs M Brelot, G Choquet et J Deny Rédacteurs F Hirsch et G Mokobodzki IV, 281 pages 1984

Vol. 1062. J. Jost, Harmonic Maps Between Surfaces. X, 133 pages 1984.

Vol 1063 Orienting Polymers Proceedings, 1983 Edited by J L. Ericksen. VII, 166 pages. 1984

Vol 1064 Probability Measures on Groups VII Proceedings, 1983 Edited by H. Heyer X, 588 pages 1984

Vol 1065· A Cuyt, Padé Approximants for Operators: Theory and Applications. IX, 138 pages 1984

Vol 1066· Numerical Analysis Proceedings, 1983. Edited by D F Griffiths. XI, 275 pages 1984

Vol. 1067: Yasuo Okuyama, Absolute Summability of Fourier Series and Orthogonal Series VI, 118 pages 1984

Vol 1068: Number Theory, Noordwijkerhout 1983. Proceedings Edited by H Jager. V, 296 pages 1984

Vol 1069· M Kreck, Bordism of Diffeomorphisms and Related Topics III, 144 pages. 1984

Vol. 1070 Interpolation Spaces and Allied Topics in Analysis Proceedings, 1983 Edited by M Cwikel and J Peetre III, 239 pages 1984.

Vol. 1071: Padé Approximation and its Applications, Bad Honnef 1983. Prodeedings Edited by H Werner and H.J. Bunger VI, 264 pages. 1984

Vol. 1072· F. Rothe, Global Solutions of Reaction-Diffusion Systems. V, 216 pages. 1984.

Vol 1073. Graph Theory, Singapore 1983. Proceedings Edited by K.M. Koh and H P Yap. XIII, 335 pages. 1984.

Vol. 1074: E W. Stredulinsky, Weighted Inequalities and Degenerate Elliptic Partial Differential Equations. III, 143 pages 1984

Vol 1075 H Majima, Asymptotic Analysis for Integrable Connections with Irregular Singular Points IX, 159 pages. 1984

Vol 1076 Infinite-Dimensional Systems. Proceedings, 1983 Edited by F. Kappel and W. Schappacher VII, 278 pages 1984

Vol 1077· Lie Group Representations III. Proceedings, 1982–1983 Edited by R Herb, R Johnson, R. Lipsman, J Rosenberg. XI, 454 pages. 1984

Vol 1078 A J E M. Janssen, P van der Steen, Integration Theory V, 224 pages 1984

Vol. 1079: W. Ruppert Compact Semitopological Semigroups. An Intrinsic Theory. V, 260 pages 1984

Vol. 1080 Probability Theory on Vector Spaces III. Proceedings, 1983 Edited by D Szynal and A Weron V, 373 pages 1984

Vol. 1081. D. Benson, Modular Representation Theory· New Trends and Methods. XI, 231 pages. 1984

Vol. 1082: C.-G Schmidt, Arithmetik Abelscher Varietaten mit komplexer Multiplikation X, 96 Seiten 1984

Vol. 1083: D Bump, Automorphic Forms on GL (3,IR). XI, 184 pages 1984.

Vol. 1084 D. Kletzing, Structure and Representations of Q-Groups VI, 290 pages 1984.

Vol. 1085: G.K. Immink, Asymptotics of Analytic Difference Equations V, 134 pages. 1984

Vol. 1086· Sensitivity of Functionals with Applications to Engineering Sciences Proceedings, 1983 Edited by V. Komkov. V, 130 pages 1984

Vol 1087 W Narkiewicz, Uniform Distribution of Sequences of Integers in Residue Classes. VIII, 125 pages 1984.

Vol 1088: A.V Kakosyan, L.B. Klebanov, J.A. Melamed, Characterization of Distributions by the Method of Intensively Monotone Operators X, 175 pages 1984

Vol. 1089 Measure Theory, Oberwolfach 1983. Proceedings Edited by D Kolzow and D. Maharam-Stone XIII, 327 pages 1984.

Vol. 1090 Differential Geometry of Submanifolds Proceedings, 1984. Edited by K. Kenmotsu VI, 132 pages. 1984.

Vol. 1091· Multifunctions and Integrands. Proceedings, 1983. Edited by G Salinetti V, 234 pages 1984.

Vol 1092 Complete Intersections Seminar, 1983 Edited by S Greco and R Strano. VII, 299 pages 1984

Vol 1093 A. Prestel, Lectures on Formally Real Fields XI, 125 pages 1984.

Vol 1094 Analyse Complexe Proceedings, 1983. Edité par E Amar, R Gay et Nguyen Thanh Van IX, 184 pages. 1984.

Vol 1095: Stochastic Analysis and Applications. Proceedings, 1983. Edited by A. Truman and D. Williams. V, 199 pages 1984.

Vol 1096 Théorie du Potentiel Proceedings, 1983 Edité par G Mokobodzki et D. Pinchon. IX, 601 pages 1984.

Vol. 1097· R M. Dudley, H Kunita, F Ledrappier, École d'Été de Probabilités de Saint-Flour XII – 1982. Edité par P L Hennequin. X, 396 pages 1984

Vol. 1098· Groups – Korea 1983. Proceedings Edited by A.C. Kim and B H. Neumann. VII, 183 pages. 1984

Vol 1099 C M. Ringel, Tame Algebras and Integral Quadratic Forms. XIII, 376 pages 1984

Vol 1100. V. Ivrii, Precise Spectral Asymptotics for Elliptic Operators Acting in Fiberings over Manifolds with Boundary V, 237 pages. 1984

Vol 1101. V Cossart, J Giraud, U Orbanz, Resolution of Surface Singularities VII, 132 pages 1984.

Vol. 1102· A. Verona, Stratified Mappings – Structure and Triangulability IX, 160 pages. 1984.

Vol 1103 Models and Sets Proceedings, Logic Colloquium, 1983, Part I Edited by G H Muller and M M. Richter VIII, 484 pages. 1984

Vol 1104 Computation and Proof Theory Proceedings, Logic Colloquium, 1983, Part II Edited by M M Richter, E. Borger, W Oberschelp, B Schinzel and W. Thomas VIII, 475 pages. 1984.

Vol 1105. Rational Approximation and Interpolation. Proceedings, 1983 Edited by P R. Graves-Morris, E.B. Saff and R.S. Varga. XII, 528 pages. 1984

Vol 1106. C.T Chong, Techniques of Admissible Recursion Theory. IX, 214 pages 1984.

Vol. 1107: Nonlinear Analysis and Optimization Proceedings, 1982 Edited by C. Vinti V, 224 pages 1984

Vol. 1108: Global Analysis – Studies and Applications I. Edited by Yu G Borisovich and Yu E. Gliklikh. V, 301 pages 1984

Vol 1109· Stochastic Aspects of Classical and Quantum Systems. Proceedings, 1983. Edited by S. Albeverio, P Combe and M Sirugue-Collin IX, 227 pages. 1985

Vol 1110. R. Jajte, Strong Limit Theorems in Non-Commutative Probability VI, 152 pages 1985

Vol 1111: Arbeitstagung Bonn 1984. Proceedings Edited by F Hirzebruch, J Schwermer and S Suter. V, 481 pages. 1985

Vol 1112: Products of Conjugacy Classes in Groups. Edited by Z Arad and M Herzog. V, 244 pages. 1985

Vol 1113 P Antosik, C. Swartz, Matrix Methods in Analysis. IV, 114 pages. 1985

Vol 1114 Zahlentheoretische Analysis Seminar Herausgegeben von E Hlawka. V, 157 Seiten 1985

Vol 1115 J. Moulin Ollagnier, Ergodic Theory and Statistical Mechanics. VI, 147 pages 1985.

Vol 1116: S Stolz, Hochzusammenhängende Mannigfaltigkeiten und ihre Rander XXIII, 134 Seiten 1985.